UNDERSTANDING ECOLOGY

UNDERSTANDING ECOLOGY

James L. Mariner

Fountain Valley School
Colorado Springs
Colorado

The Independent School Press

WELLESLEY HILLS
MASSACHUSETTS

PRINTED IN THE UNITED STATES OF AMERICA.

0-88334-070-4

To
BROWNIE

CONTENTS

PREFACE

The increasing concern with which we have been forced to view the environment is not limited to pollution and its effects. Indeed, we cannot begin to understand the effects of man's activities upon the environment without first understanding how the natural environment operates. The purpose of this book is to introduce you to the various components of your surroundings and to investigate the ways that they interact to produce the situations that you might observe.

Most of you will not be scientists, and fewer will be biologists. Nevertheless, it is extremely important that all of us understand and come to terms with our environment. It is not a matter for scientists alone. Rather, laymen, legislators, businessmen, and in fact, all of us make daily decisions that affect the environment in some way, no matter how small. Therefore, we need to understand it thoroughly because our very survival may depend upon it.

Beyond the necessity of understanding the operation of our environment, it would seem desirable that we develop an appreciation for it and how it came to be the way that it is. The natural world did not just happen. The interaction of innumerable forces has resulted in a tremendous variety of life forms, each uniquely adapted to its particular way of life. To understand and appreciate the results of these complex forces is to gain a new relationship with them. The intricate interplay that exists between various members of the living world as well as that existing between these members and the non-living world is truly mind-boggling. To this end, I would hope that various ecological phenomena discussed here will have a lasting impact so that even as a vacation traveler, you will be aware of the kinds of things you will see and will understand and appreciate how they came to be that way.

Except for occasional references in this book, the study of man's impact upon the environment, the reduction in the quality of life that results from pollution and overcrowding, and the solutions to these problems are beyond the scope of this book. This material is being constantly updated, and there is an abundance of current literature available on these and related subjects. Ecology, as a study in itself, is usually limited to a single chapter or two in the secondary level biology textbooks, and is often left to the professional biologist. This text is an attempt to fill the void.

I would like to extend my sincere appreciation to those who have been instrumental in the development of this book: to F. Martin Brown, without whose encouragement and stimulation this book would never have been written; to Robert Peters, a student and friend, for his artistic touch; to Thelma Goodnight for her perseverance in the typing of the original manuscript; and especially to my students, who suffered through the initial phases and constructively offered their invaluable criticisms.

January 1975 JLM

UNDERSTANDING ECOLOGY

1

"We abuse the land because we regard it as a commodity belonging to us. When we see the land as a community to which we belong, we may begin to use it with love and respect."

*Leopold**

Introduction to Ecology

There are few among us who, at some time or another have not turned over a stone and discovered the array of life which suddenly presented itself. It probably didn't occur to you at the time, but you had just uncovered a tiny corner of the life which exists all around us, although usually unseen. Here was a special community, self-dependent and operating within a delicate balance that ensures the perpetuation of that community. Had you stopped to observe the hustle and bustle going on in front of you, you might have seen several types of insects, a spider or two, some "pill" bugs, an earthworm, or perhaps even a snake or a lizard. Such an aggregation would not be uncommon. Many members of this tiny community would probably go unnoticed. Several soil organisms, including bacteria and other microscopic organisms could not have been detected without the aid of special techniques.

Why do these organisms live in such a place? How are they able to do so? What factors contribute to making a community just right for the conditions which their particular pattern of life demands? And at least equally important, what has all of this got to do with me? These are perhaps some of the questions which might occur to you as you reflect over such a community. Certainly this world would be a quite different place if such organisms did not exist. Whether you realize it or not, directly or indirectly, such communities exert their influence upon you and your life. And likewise, you exert an influ-

*Leopold, Aldo. 1949. *A Sand County Almanac.* Oxford University Press, London, p viii.

ence upon such communities. The turning of the stone was itself a drastic alteration of the conditions which existed before the stone was turned.

The fact that man can exert a profound influence upon delicately balanced communities, including his own, is proof that we must obtain a thorough understanding and appreciation of the factors which govern the operation of communities. The turning of a stone is relatively minor compared to the changes man can effect by cultivating the soil, polluting the environment with poisons and sewage, and introducing radioactivity from explosions and reactors.

Biology is the study of life, and although this is a relatively simple definition, the numerous facets of this study reveal the truly complex nature of life (Fig. 1-1). To study each of the areas involved in a total survey of biology would be impossible in a single year, and one would soon lose the concept of the forest in the study of the trees.

Few of you will be scientists; fewer still will be biologists. But one does not have to be a biologist to appreciate many of the aspects of life. You are a living organism yourself, and you are constantly surrounded by other forms of life, similar in many respects to you. Therefore, it is only logical that a study of life begin with life as it exists every day in order to see how it involves us as human beings. Beginning here, we embark upon the study of life through the facet of ecology.

Just as in any other field of study, biology has its levels of organization. Perhaps you can best visualize the relationships between the various levels if you think of the study of life as a biological spectrum in the sense that visual light is a spectrum. The colors have no discrete lines dividing them, and one blends into the next presenting a rainbow of continuous color. The spectrum of biology may be conveniently divided into ten levels of organization with each level lacking definite lines of separation from adjacent levels (Fig. 1-2). Ecology concerns itself with the six levels at the right of the spectrum, and it is with these areas that we shall primarily concern ourselves in our pursuit of biology.

Ecology is the study of the interrelationships existing between the environment and living organisms. Organisms cannot exist by themselves. They are dependent upon many features

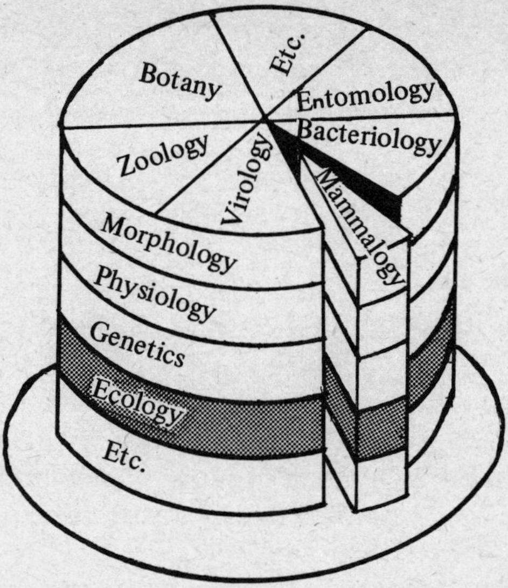

Fig. 1-1. The biology "layer cake." The horizontal layers represent some of the many facets of biology while the sections of the cake represent various taxonomic areas of biology which deal with restricted organisms or types of organisms. Note that a complete study of a given type of organism involves all of the layers, and that each layer involves every section. (Adapted from Odum, 1953.)

BIOLOGY

P	C	T	O	O	P	C	E	B	B
R	E	I	R	R	O	O	C	I	I
O	L	S	G	G	P	M	O	O	O
T	L	S	A	A	U	M	S	M	S
O	S	U	N	N	L	U	Y	E	P
P		E	S	I	A	N	S	S	H
L		S		S	T	I	T		E
A				M	I	T	E		R
S				S	O	I	M		E
M					N	E	S		
					S	S			

ECOLOGY

Fig. 1-2. The field of biology as a spectrum of levels of organization.

in their environment, and in turn, serve a function in that environment which enables other organisms to survive. A rabbit, for example, is dependent upon certain *abiotic* (non-living) features such as the availability of water, the temperature of the environment, oxygen in the air, and a host of other factors, either directly or indirectly. He is also dependent upon the *biotic* factors of his environment such as certain types of plants which he can use for food and perhaps others which he can use for shelter. He, in turn, is depended upon as food for other organisms such as a fox or lynx, a passing hawk or owl, and many unseen organisms which serve as parasites within or upon his body. In a rather discrete manner, the rabbit is also dependent upon the fox for controlling his population. Without this check existing on the growth of rabbit populations, they would soon literally eat themselves out of house and home with drastic consequences for the rabbit!

By his activity, the rabbit exerts a powerful influence upon his environment. His grazing continually crops the grasses of the community, and when food is scarce, small trees are stripped of the tender bark around their bases, stunting their growth or even killing them. Wounded trees are susceptible to invasion by disease or insects which may cause further damage to the community. Thus, the rabbit is not alone in a community; rabbits are dependent upon and are depended upon by other members of the community in which they live. Their presence does not go unnoticed by the community.

Although a study of ecology might begin with a detailed account of an organism, one quickly realizes that the organism itself is only a part of a functioning unit, the *population.* As ecologists, we understand that a population is a group of organisms, similar in kind, living in a particular area at a given time. Certainly the effect of a population of rabbits living in a certain environment is much greater than that of a single rabbit. The effect will vary from year to year, also, since the population does not remain constant for a variety of reasons. Food supply, migration, and disease are only three of the many reasons why populations tend to fluctuate from year to year.

All of the populations living in a given area comprise a *community.* Just as a community of human beings might be made up of doctors, lawyers, grocers, merchants, teachers and many others, a natural community is composed of many different populations. For example, a forest community might contain oak trees, hickory trees, gooseberry bushes, song sparrows, blue jays, red foxes, opossums, terrapins, insects of many descriptions, and a variety of others. A community implies some degree of interaction as well. While it may not be totally independent, many intimate relationships exist within a biotic community as populations interact with each other.

As we have noted above, the environment is not composed exclusively of biotic factors, but includes non-living features as well. Thus, the community is surrounded by abiotic features, and together, they make up a functional unit, the *ecosystem.* Generally speaking, the ecosystem is an independent unit. Plants supply the food for the animals, and decaying animal and plant bodies provide nutrients for the plants.

Another term that is often used in place of ecosystem is the term *habitat.* It is at the ecosystem level that ecology mostly concerns itself. In fact, ecology has sometimes been called habitat biology.

Ecosystems come in all shapes and sizes, from large expanses of grassland to a few square millimeters on an exposed rock, from a mountain pine forest to a coral reef in the ocean. Some are large enough to include many smaller ones. Although two ecosystems may be similar with respect to the conditions existing within them, there will, nevertheless, be differences between them. In addition, within a given ecosystem, regardless of size, conditions will vary from place to place. In the grassland habitat mentioned above, for example, there will be variations in temperature, humidity, wind velocity, sunlight and available moisture depending upon whether conditions are measured at ground level, a foot or two above the ground, on a hill or in a draw. But such variations do not complete the picture, for there are seasonal variations as well. Related to climatic factors existing in an ecosystem over a period of a year, biotic factors vary also. Birds and insects may migrate into and out of the area, vegetation may become dormant during winter or dry seasons, different forms of insects will occur at various times throughout the growing season, and even microorganisms will show wide variations in kinds and numbers at different seasons.

Plants may adapt to the harsh conditions of winter by becoming dormant. The previous season's growth dies and accumulates on the surface of the ground, insulating it from the cold.

Extremely large-scale ecosystems may make up a *biome.* Several types of ecosystems of varying sizes may exist in an area that has the same general over-all climatic conditions, vegetation, and animal life. The grassland, again, is not all just grass and grazing type animals. It is also dissected by rivers along whose banks for sometimes as much as a quarter of a mile on either side are found almost impenetrable tangles of trees, shrubs, vines, and a host of herbs not found out in the grassland area itself. Climatological variations from one extreme end of a grassland area to another may cause variations in the types of grasses and the animals living there. The vast expanse of the prairies, extending from Iowa, Missouri, Kansas, Nebraska, Oklahoma, Texas, and the Dakotas to the Rocky Mountains is a prime example of variation within a biome. In the eastern portion of the prairies, where moisture is more abundant and the soils permit, bluestem, Indiangrass and brome abound. But up against the east slope of the Rockies, where the soil is more sandy and rainfall is less than one-third of that on the eastern sections, the short-grass prairie exhibits buffalo grass and a variety of gramas. A more detailed study of the biome concept and specific biomes will be taken up in Chapter Seven.

Fig. 1-3. Ecology has far-reaching applications in many areas that affect our lives. (Diagram from Hylander, 1966.)

That portion of the earth that supports life, and includes all of the biomes is called the *biosphere.* To properly understand the concept of the biosphere, you must picture it not as a "thing," but as a "system." Such a system is actually an assemblage of all of the biotic and abiotic parts of the earth that combine to form an interrelated unit. The biosphere, then, involves not only the living things, but the relationships that exist between them as well as the materials and the required energy that supports them.

The ecologist must not only have a good foundation in the necessities of life, but he must thoroughly understand how these necessities are made available to all of the organisms in an ecosystem. He is interested in how the ecosystem maintains the delicate balance that is required for its operation and the constant bombardment of the ecosystem with upsetting influences. In this day and age of exploitation of the environment by man, he is particularly concerned with the effects of pollution, uninhibited population growth, and the removal of large areas of land from productive use. The remainder of this book will concern itself with analyzing the important features of an ecosystem and determining their importance and relationship to man.

QUESTIONS FOR DISCUSSION

1. What is meant by the terms biotic and abiotic? Give some examples of each.
2. In what unique way might a deer population be dependent upon a predator such as a mountain lion?
3. Define an ecosystem, and explain how it may be considered an independent unit.
4. Make a distinction between the concepts of population and community. Give examples of each.
5. What is meant by the term biosphere?

INVESTIGATIONS AND ACTIVITIES

1. Collect a variety of newspaper and magazine articles that reveal problems confronting man in his environment. Abstract each article and explain how each applies itself to the problem of the ecology of man.
2. As Charles Darwin discovered, in his visit to the Galapagos Islands, and other scientists have more recently discovered, certain islands present excellent opportunities to study native populations of organisms. One such living laboratory has been extensively studied by biologists Durward Allen and David Mech. An account of their work appears in the February, 1963 issue of *National Geographic.* Make a written report on the interaction of the moose and timber wolf populations on Isle Royale as studied by these two biologists.

RECOMMENDED READING

Allen, Durward L., and L. David Mech. 1963. *Wolves versus Moose on Isle Royale.* Nat. Geogr. 123(2): 200-219.

Bates, Marston. 1964. *Man In Nature*, 2nd ed. Prentice-Hall, Inc., Englewood Cliffs, N.J.

Cole, LaMont C. 1958. *The Ecosphere.* Sci. Amer. 198(4): 83-92.

Hylander, Clarence J. 1966. *Wildlife Communities from the Tundra to the Tropics in North America.* Houghton Mifflin Co., Houston.

Odum, Eugene P. and Howard T. Odum. 1959. *Fundamentals of Ecology,* 2nd ed. W. B. Saunders Co., Philadelphia.

2

". . . . When we regard every production of nature as one which has had a long history. . . . how far more interesting does the study of natural history become."

Darwin*

The Web of Life

COMPONENTS OF AN ECOSYSTEM

We have seen how the various parts of the ecosystem exert profound effects upon each other. In order to understand exactly how these components interact, it is convenient to divide the ecosystem into four broad categories: (1) producers, (2) consumers, (3) decomposers, and (4) abiotic substances.

Producers are those organisms which are able to manufacture food from simple inorganic substances. Since they are continually replacing food used by other organisms, producers represent the foundation of the living world and are thus the first link in a food chain. They are usually green plants and because they are capable of producing their own food, they are said to be *autotrophic.* Producers rank in size from large trees and shrubs, to grasses and other small herbs, on down to microscopic, waterdwelling plants called *phytoplankton.*

Consumers are *heterotrophic* organisms; that is, they cannot produce their own food. They are generally animals and as a group, make up the next links in a food chain. Two general distinctions may be made among consumers on the basis of their feeding habits. *Herbivores* feed directly on the producers, converting some of the plant material into animal

*Darwin, Charles. 1859. *The Origin of Species.* Random House, New York. p. 371.

tissue. *Carnivores* obtain producer material also, but only indirectly. They rely directly upon other heterotrophs for their food materials, adding a degree of complexity to the food chain. You will quickly realize that a third alternative is possible. There are also some animals, including man, which feed as herbivores at one time and as carnivores at another. These animals are called *omnivores.* Such organisms add still greater complexity to a food chain. Organisms which obtain their food in the same number of steps, beginning with plants, are said to belong to the same trophic level. Thus, green plants occupy the first trophic level, herbivores a second trophic level, and carnivores, second, third, or even fourth levels.

A naturally-occurring food chain in the ocean might begin with small phytoplankton. These would be eaten by shrimp, for example, which are in turn eaten by certain fish. Some of these fish may be eaten by seals and some seals may be prayed upon by killer whales. Figure 2-1 summarizes such a food chain.

However, if we were to analyze all of the food taken in by each of the organisms just described, we would invariably discover that such a simple, direct chain does not exist. It is far more likely that there are other organisms involved, being linked to the original chain at several points. Such complex relationships begin to take on more the characteristics of a *food web* rather than that of a chain.

Decomposers make up a third important element of the ecosystem. These organisms, chiefly bacteria and fungi, have the task of transforming dead organic material back into simpler raw materials which can again be utilized by the producers.

The final essential components of a functioning ecosystem are the abiotic substances. Many of these substances are made available by decomposers and other organisms. Others come from the soil itself. Such substances include a variety of chemicals, such as calcium, nitrogen, phosphorus, water, oxygen, carbon dioxide and a host of others.

The mountain lion is an example of a carnivore, relying upon other heterotrophs for its food supply.

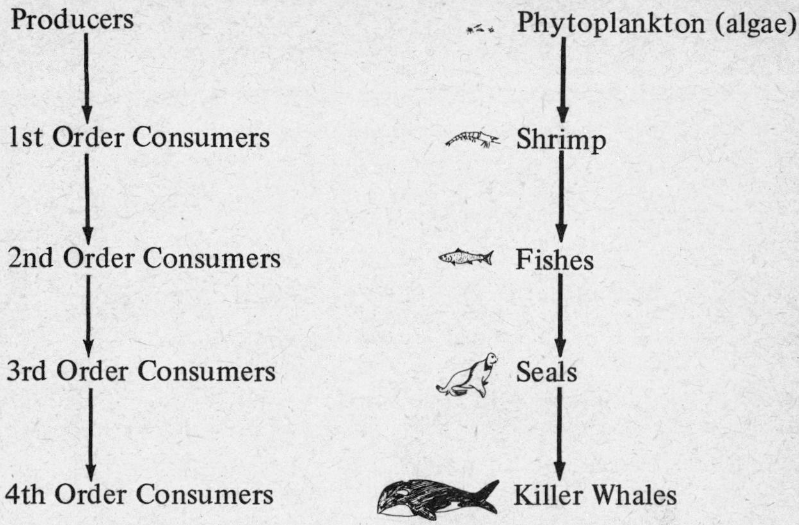

Fig. 2-1. Summary of a typical organic food chain.

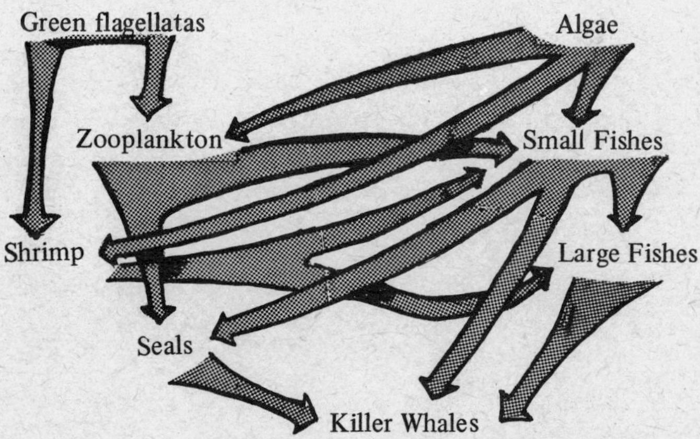

Fig. 2-2. A simplified food web.

The mule deer is a first order consumer feeding directly from the producer trophic level.

The idea has been presented in Chapter 1 that an ecosystem may be of any size, from a few square millimeters to a large expanse of ocean. The essential point that should be well understood is that an ecosystem must contain the four basic components and that they operate together as a functional unit, even if for only a short time. Upsetting influences are constantly making themselves felt in nearly all ecosystems, but usually, a well balanced system is capable of re-adjusting itself to attain a balanced state. Thus, the equilibrium that is established is a dynamic one, constantly shifting to alleviate stresses which are constantly arising.

ECOLOGICAL NICHE

The functional role of an organism within a community is called the *ecological niche.* It refers to what the organism does and how it fits into the overall operation of the community. It is perhaps unfortunate that the term niche is used in this context since a dictionary definition would describe it as a place. The ecologist, however, uses the term habitat to describe where an organism lives, and the term niche to describe how it lives. A rough analogy at this point might serve to prevent problems in the use of these terms. The habitat of an organism may be thought of as its "address"; its niche would be its "profession."

The producer fills perhaps the most essential niche of all organisms. Without the producers, the heterotrophs could not survive, and life in a community would soon come to an end. It is obvious that the producers are not the same in all communities, nor will all producers in a community be of equal importance. In a mature forest community, for example, the important producers will be a variety of trees and a few shrubs that will survive in the shade of the trees. There may also be other less important members of the producer group in the form of grasses, vines, and other small shrubs and herbs. In a prairie community, grasses will be the important form of producer organisms with some small herbs filling a somewhat lesser niche. Producer organisms, in addition, may fill other equally important niches in a community. They may serve as shelter for birds and other animals. They may also modify the environment slightly to enable other organisms to survive in the community. For example, shade allows certain organisms to survive better than they would in more direct sunlight. Producers also shelter lower levels by providing an insulating blanket over the land, reducing temperature changes and retaining moisture.

The herbivores serve as a food source for the carnivores. They obtain their food by grazing the producers and converting plant tissue into animal tissue. Herbivores also differ with respect to the type of community in which they are found. Pronghorn antelope, rabbits and certain species of field mice will be common in a grassland community. A forest community, on the other hand, might sport elk, deer and other species of mice. The carnivores fill their niches by controlling populations of herbivores and lower level carnivores. They, too, are respectors of the environment, and the particular species encountered will depend upon the community. The polar bear will be found in the Arctic tundra; a coyote is common in the grasslands and the desert.

The niche of the decomposer has already been described briefly. It is to convert dead organic matter back into simple raw materials which can be utilized by the producers. Thus, the cycle is complete: raw materials, converted into organic compounds by the producers and passed along to various levels of consumer organisms are finally returned to the soil as raw materials by bacteria and fungi as they fulfill their niche as decomposers.

Many organisms are extremely specialized in fulfilling their niches, and will die out if conditions are not ideally suited to their way of life. Certain insects are able to live only if a particular type of shrub is available as a food source for the young larvae. Other organisms exhibit a wide diversity in adapting to an environment, and are able to fill a variety of niches depending upon the particular conditions which exist at that time. Man is a good example of an organism with a wide niche diversity. But whoever may occupy a niche at a particular time, does not possess a permanent lease because his permanence is subject to a variety of factors, making his presence rather insecure. Just because he is well adapted to his situation is no guarantee that another organism is not better adapted to fill the same niche. It is not uncommon that an invading species is capable of successfully competing with and displacing an existing species. Man has often been guilty of transporting aggressive species to new areas only to have the introduced species completely displace existing, more desirable organisms. Man may introduce organisms into new areas accidentally or intentionally. Sometimes the results are extremely beneficial as witnessed by the cultivation of many grain and citrus crops introduced from distant countries. But often, the results have been disastrous. The release of rabbits into Australia resulted in their rapid dispersal throughout much of the continent with consequent ravaging of large areas of grazing land. With no natural predators to control their numbers, they were able to multiply at a fantastic rate and compete successfully in the new environment. The introduction, in 1890, of the starling into America is another example where an introduced species has come to exist in pest proportions over most of the United States. Rats, mice, and bedbugs are other examples that have been introduced from Europe although not intentionally. Many organisms introduced into new areas are not able to survive without constant cultivating. But many countries and states have established inspection and quarantine stations at their borders to prevent accidental or intentional introduction of potential pests into new areas.

And so it is that when two separate species that occupy the same niche come into contact, they become competitors for food and space. That species which possesses the best adaptations for successfully competing with other organisms will be able to become the dominant organism for that niche. The others will not survive there under normal conditions, or at best, their numbers will be kept to a bare minimum in that area unless they possess a wide range of adaptability and can occupy another more or less vacant niche in that area.

On the other hand, two distinct species can fill the same ecological niche if they are not in competition. Usually, this is the case with organisms located on different continents. Widely separated areas of similar ecological makeup will very likely be made up of different species of organisms. Nevertheless, the niches are similar and are filled by *ecological equivalents*; that is, by different organisms that perform similar functions in different (but similar) communities with respect to habits, food and relationship to other organisms. For example, the pronghorn fills a niche in North America that is equivalent to the wild horse in Eurasia, the antelope and zebra in Africa, and the large kangaroos in Australia. Other examples are given in Table 2-1.

ENVIRONMENTAL CYCLES

To say that each biotic component in an ecosystem is dependent upon another component is only telling half of the story. While it is true that the consumers are dependent upon the producers and the producers in turn depend upon the available abiotic materials of the soil and atmosphere, the story is incomplete for a very important reason. To build complex organic substances from relatively simple raw materials requires vast amounts of energy. Unless energy is supplied in abundant quantities, the system begins to run down. It is precisely this reason that a perpetual motion machine cannot be built – there must always be an outside energy source to keep it moving.

And an ecosystem is no different. It, too, needs an outside energy source. In this case, the source is the sun. Green plants have evolved a special component which enables them to momentarily utilize some of the radiant energy from the sun long enough to convert that energy into another form – chemical energy. This complex process is known as *photosynthesis.*

The first step in the production of organic materials by photosynthesis is the manufacture of simple sugars. These relatively simple substances are then able to furnish the energy necessary for additional steps in food production. Starches and fats may be constructed for energy storage, and proteins may be synthesized in the growing process. All of these organic substances have in common the elements carbon, hydrogen and oxygen.

Table 2-1. Ecologically equivalent organisms in terms of feeding habits or stratal location representing several geographic areas of the world.*

	Geographical Areas			
Types of Animals	Eurasia	Africa	N. America	S. America
Mammals				
Herbivores	Red deer	Fallow deer	Wapiti (elk)	Bracket deer
Carnivores	Leopard cat	Golden-haired cat	Bobcat	Jaguarondi
Birds				
Frugivores	——	African gray parrot	Military macaw	Yellow-headed amazon
Carnivores	Griffon vulture	Milky eagle owl	Screech owl	Harpy eagle
Reptiles				
Ground snakes	Common cobra	Black-necked cobra	Prairie rattler	Cyclatras
Tree snakes	Golden tree snake	Black tree snake	Rough green snake	Green tree boa
Insects				
Locusts	Migratory locust	Desert locust	Rocky Mountain locust	S. American locust

*Adapted from Knight, 1965.

In addition, all proteins possess the element nitrogen. Consequently, it becomes obvious that these elements as well as a few others, are necessary for life as we know it to exist, and therefore must be in constant supply.

It is especially interesting to note that only green plants can make use of the raw materials necessary for synthesizing the complex organic compounds necessary for life. Furthermore, although carbon and nitrogen exist in many forms in the environment, there is but one form of each that green plants are capable of utilizing in the production of food. Carbon dioxide found in small amounts in the atmosphere is the source of all organic carbon. In this form, it is readily incorporated into simple sugars in photosynthesis. Although some 78 percent of the atmosphere is composed of nitrogen, it is unavailable to most plants for the synthesis of protein. Nitrogen must be combined with oxygen and made available as a nitrate.

Since carbon and nitrogen are so crucial to life and their availability is dependent upon constant supplies of carbon dioxide and nitrate compounds, it becomes of ecological importance to understand how these raw materials are constantly made available to the green plant. Conveniently enough, nature has evolved built-in mechanisms for the constant replenishment of the necessary materials. Each of these elements is involved in a cycle whereby it is used, stored temporarily in the bodies of plants and animals, and eventually released for use again.

Carbon Cycle. Green plants are constantly removing carbon dioxide from the atmosphere by photosynthesis. The carbon compounds thus formed are in turn incorporated into the bodies of animals. As populations of plants and animals increase, more and more carbon dioxide is being removed from the atmosphere. Some of it is returned directly to the air by plant and animal respiration. In this process, the complex organic substances are constantly being broken down into the simple materials from which they were originally synthesized. The purpose for such breakdown is to provide the organism with energy for continued growth, warmth, and a myriad of other chemical activities needed to carry on life. The result is that carbon dioxide, along with other waste materials, is released.

But, if plant and animal respiration was the only source of carbon dioxide, it would not be long until the total supply would be completely used up with the subsequent disappearance of life from the earth. It is estimated that the present level of carbon dioxide in the air, some 0.03% of the atmosphere, would support the present plant population for less than 50 years! What, then, is the source of the carbon dioxide that can supply the necessary quantities of this vital raw material? The answer lies in the decay organisms, the decomposers which are the third biotic component of an ecosystem. Without these small but important organisms, the world's supply of carbon dioxide would soon be exhausted. In addition, the present world would not be a pleasant sight if the surface was littered by the dead remains of all of the organisms which had ever lived. Thus, the service of the decomposers is twofold: they remove the remains of the past and at the same time provide the vital link in a cycle allowing the perpetuation of life. Figure 2-3 summarizes the essential parts of the carbon cycle.

Nitrogen Cycle. The decomposers act in another important capacity. It has already been pointed out that the air cannot ordinarily be utilized as a source of nitrogen for organisms, and that nitrate compounds found in the soil constitute by far the largest source of nitrogen for protein synthesis. Just as in the situation for carbon, nitrogen is continually being incorporated into plant and animal tissues and its availability would soon be reduced to nothing if it were not for the decomposers. A variety of bacteria, living in the soil, are available to carry out a series of steps in breaking down nitrogen-containing compounds found in the waste materials of animals and the dead bodies of all organisms.

Some bacteria are able to break down proteins into ammonia, a relatively simple nitrogen compound. Other bacteria possess the ability to convert ammonia into still simpler nitrites. Still other bacteria convert the nitrites into nitrates which are then available in the soil to be again taken up by plants and utilized in the synthesis of protein. Nitrates may also be made available as certain types of rock particles dissolve or by the addition of fertilizers by man. Another important source of nitrates is the air itself. Some forms of bacteria and blue-green algae are able

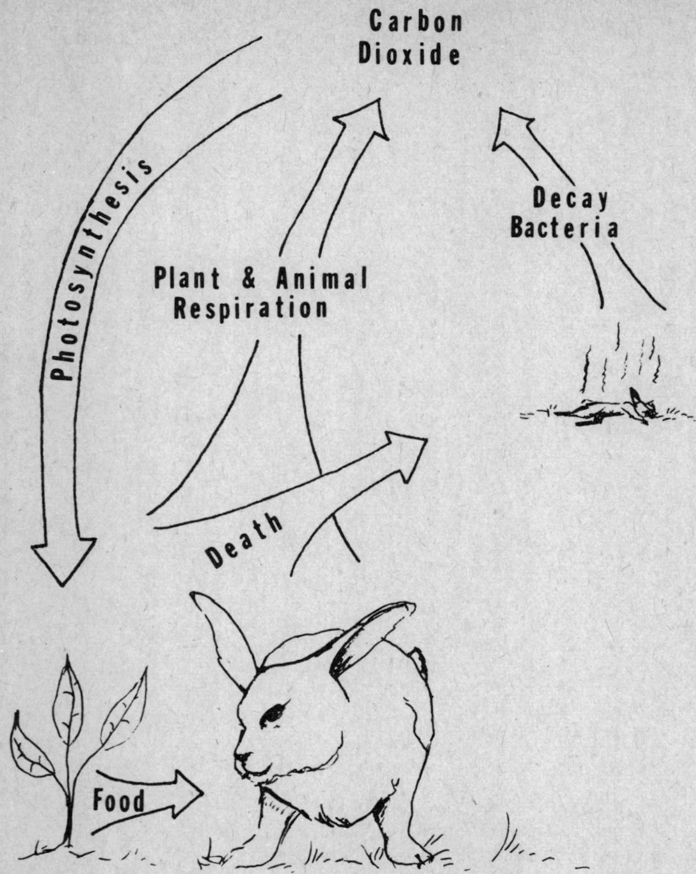

Fig. 2-3. The essential steps of the carbon cycle.

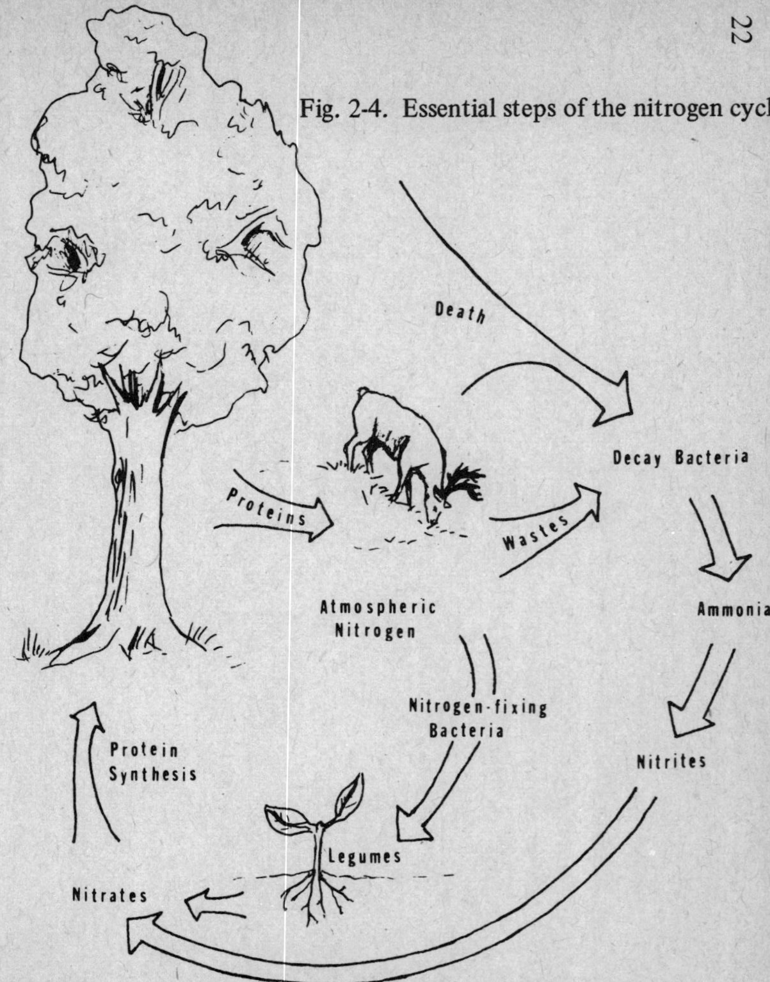

Fig. 2-4. Essential steps of the nitrogen cycle.

to take in nitrogen from the atmosphere and incorporate it into structural proteins. Some of these organisms are free-living; when they die, their protein is made available to the nitrifying bacteria mentioned above. Other bacteria live as *symbionts* in modules on the roots of legume plants. The nitrogen "fixed" into compounds by these so-called "nitrogen-fixing bacteria" becomes available as usable nitrogen and is more than enough for the plant on which the bacteria are living. The remainder is deposited in the soil to be used by other plants (Fig. 2-4).

Naturally, the producers, consumers and decomposers are not aware of the grand scheme of things involved in these cycles. The presence of life and its inevitable production of waste material has led to the consequent evolution of organisms which are capable of utilizing these waste materials as an energy source. The use of these substances, according to physical and chemical laws also inevitably leads to the low energy raw materials which brings the cycle back to its starting point.

Life as we know it is dependent upon many additional substances, two of which are often underestimated but have been alluded to previously: water and oxygen. Particularly in this modern age of an affluent society (often perhaps appropriately dubbed as an effluent society), the importance of these two substances begins to take on a new meaning. Until recently, the availability of these materials was a fact of life and taken for granted. But today, increasing pressures are being placed upon governments to produce legislation that would release funds for the development of new fresh water sources and anti-pollution devices which would curtail the destruction of our water and air.

Two-thirds of the surface of the globe is water. It is continually evaporating into the atmosphere, returning to the earth as rain or snow. Some of this precipitation returns directly to the sea. That which falls onto the land may follow a devious path before it completes the journey to the sea, but that is its ultimate goal. It may be absorbed into the soil to be taken up by plants and either again evaporated into the atmosphere from the leaves or be split into its component elements of hydrogen and oxygen by photosynthesis. The oxygen

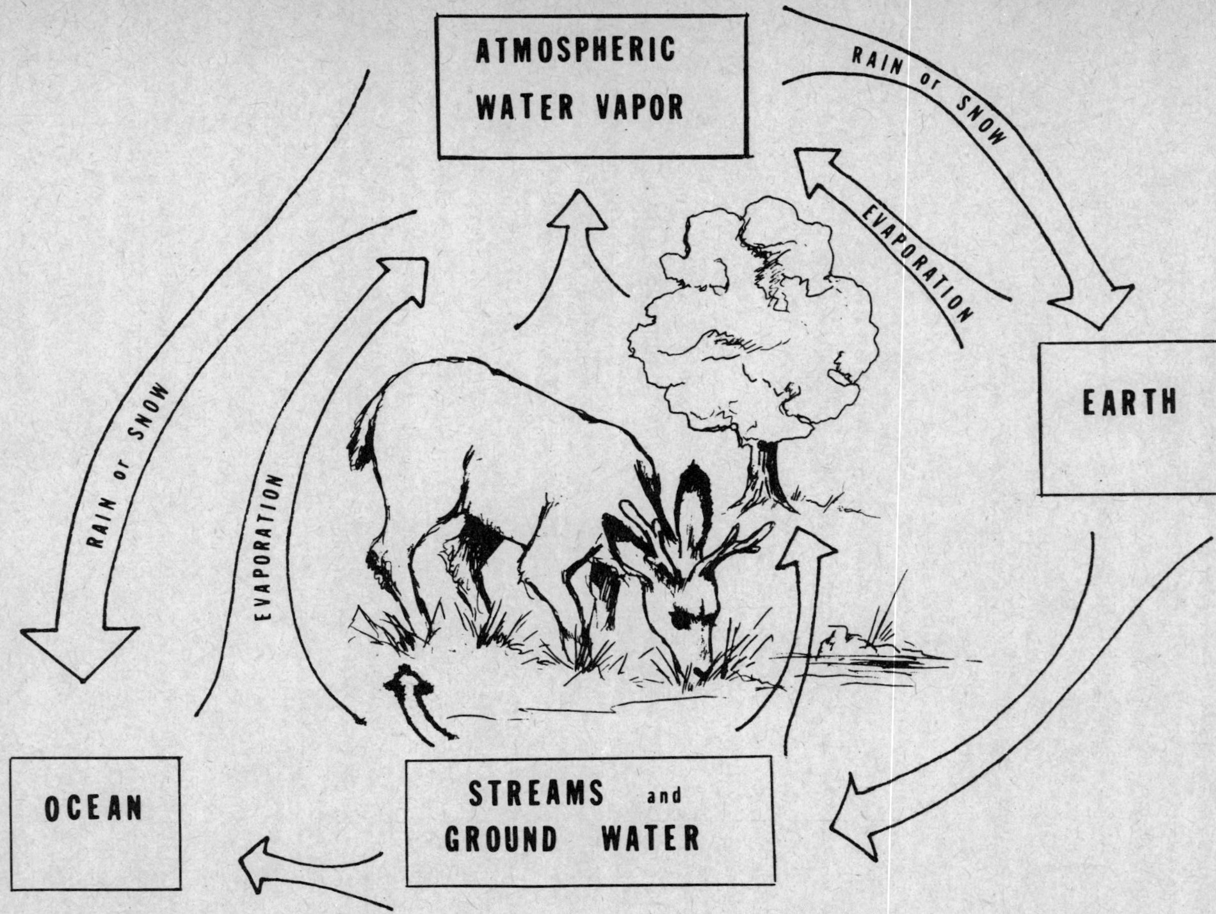

Fig. 2-5. The water cycle touches every facet of life.

is thus made available to plants and animals for respiration. Some of the water may run into streams, lakes and rivers coursing always to lower elevations and the waiting sea. Along the way some will be used by animals and plants, but only temporarily as it eventually is again deposited and picked up by another agent unwittingly helping to complete the ancient cycle (Fig. 2-5).

The oxygen released in the photosynthetic process makes up approximately 20 percent of the atmosphere. Without it, both the carbon and water cycles would be incomplete because it forms an indirect link between these two essential cycles. Originating as a chemical part of water, it ends its journey in the same form. It is essential to the respiratory process in the carbon cycle, for without it, complete energy release and consequent release of carbon dioxide would be severely hampered.

There are a great many such cycles involved in circulating materials through the environment. While they are of perhaps lesser importance, their significance should not be overlooked. The four cycles presented here serve as examples of how they operate and are able to continue to do so. They are presented also to illustrate the intricate balance which exists in nature's economics of supply and demand. Interruption of the economy can produce serious effects throughout the system, and man has proven to be a serious threat to this intricate economy. Although the system possesses a truly amazing ability to "bounce back" after being interrupted, there is no guarantee that this will continue indefinitely with continued increase of interruptions.

The dumping of sewage into streams greatly alters the natural balance present in aquatic ecosystems, and threatens the availability of a valuable resource.

Tons of pollutants are poured into the atmosphere daily by automobiles, factories and a variety of other sources.

Particulate matter and other gaseous products accumulate in the air and undergo a variety of chemical reactions creating smog that is irritating to the skin and membranes of the eyes and lungs, and damaging to the health of plants and animals as well as property.

QUESTIONS FOR DISCUSSION

1. Explain the relationships which exist between producers, consumers, and decomposers.
2. Why is the concept of a food web more descriptive than that of a food chain?
3. In terms of ecology, what is meant by the term dynamic equilibrium?
4. Show by description that you understand the concept of niche.
5. What is the principal process occurring in green plants that provides the living world with an energy source? What is the ultimate source of the energy for the living world?
6. What fundamental process enables plants and animals to release and utilize the energy provided by green plants?
7. Cite some examples of organisms which are capable of occupying more than one trophic level.
8. What two fundamental "services" are provided by the decomposers?
9. Outline the essential steps involved in the four basic environmental cycles.
10. Describe how the producers affect a community both directly and indirectly.
11. What do you think would happen to the "balance of nature" in a pond community if one kind of organism suddenly increased in numbers? Would it make a difference if the expanding population were a green algae (a producer) or a fish (a consumer)?
12. What is the relationship between the fossil fuels used today by man in industry and home, and the fuels with which he feeds his body?

INVESTIGATIONS AND ACTIVITIES

1. Using a simple chlorophyll-containing organism such as Elodea (Anacharis) or other alga, investigate the relationship which might exist between the number of chloroplasts per cell and amount (intensity) and/or color of light which the cells receive.

2. Diagram a food web that might exist in a community which you have observed recently. Present evidence to show that your diagram is reasonable.
3. It has been pointed out in the chapter that at least theoretically no two species occupying the same ecological niche live together in the same habitat (except possibly temporarily). Using at least two closely related species found in your area, study their respective niches in an attempt to demonstrate support for this theory. (Additional information for a study of this type may be found in Research Problems in Biology, Series 2, pp. 129-132. Anchor Doubleday.)
4. Demonstrate the relationship between producers and consumers by preparing seven tubes each half-full of distilled water to which have been added 2-3 drops of a bromthymol blue indicator. The effect of adding carbonated water to one, blowing for a minute or two through a glass tube into another, adding a weak acid to a third, a sprig of Elodea to a fourth, a snail to a fifth, a sprig of Elodea and a snail to a sixth, and using the seventh as a control should yield some valuable observations to illustrate the interdependence concept. Write up your results in a standard report form.
5. Commercial jetliners are a constant source of carbon dioxide and water vapor for the atmosphere, especially in large metropolitan areas. However, their release at extremely high altitudes poses a special potential problem for the biosphere. Write a brief summary of the importance of these waste materials to the biosphere as they relate to high altitude release.

RECOMMENDED READING

Ambroggi, Robert P. 1966. *Water under the Sahara.* Sci. Amer. 214(5): 21-29.

Avery, G.S. 1957. *The Dying Oak.* Sci. Amer. 196(5): 112-122.

Cole, LaMont C. 1958. *The Ecosphere.* Sci. Amer. 198(4): 83-92.

Hylander, Clarence J. 1966. *Wildlife Communities from the Tundra to the Tropics in North America.* Houghton Mifflin Co., Houston.

Leopold, Aldo. 1949. *Odyssey.* In: *A Sand County Almanac.* Oxford University Press, New York.

Kriebs, Jean Oak and Albert Schatz. 1973. *Who Ate Roger Williams?* Amer. Biol. Teacher. 35(3): 155.

Leshan, E., *et al.* 1965. *Sabine Grove Ecology Study.* Natur. Hist. 74(5): 14-23.

Odum, Eugene P., and Howard T. Odum. 1959. *Fundamentals of Ecology,* 2nd ed. W. B. Saunders Co., Philadelphia.

Simpson, G.G. and W. S. Beck. 1965. *Life: An Introduction to Biology,* 2nd ed. Harcourt, Brace & World, Inc., New York.

Woodwell, George M. 1967. *Toxic Substances and Ecological Cycles.* Sci. Amer. 216(3): 24-31.

3

"New molecules have come from the grass and the bodies of animals to be part of me a little while . . ."
Eiseley*

Energy in the Ecosystem

PRINCIPLES OF PHYSICS

All living organisms in the ecosystem require energy to continue the activities of life. Each organism is in the continual process of obtaining its energy in the form of food by feeding upon the organisms of lower trophic levels. The usually complex nature of a food web as it occurs in nature has already been described.

But there is a limit to the amount of energy that can be made available in the biosphere. Green plants are capable of utilizing for photosynthesis only about one-half of the radiant energy reaching the earth from the sun. The energy which is used by photosynthesis directly is converted into a storage form – chemical energy. Part of the remainder is reflected back into the atmosphere, and some is absorbed by the plant and converted into heat. As heat, this energy is not entirely lost, however. Even though it cannot be used in the photosynthetic process directly, it is still used to circulate materials within the plant. It aids in the evaporation of water from the leaves, a process known as *transpiration.* The process is essential because plants possess no organ to circulate materials as do most animals.

*Eiseley, Loren, 1957. *The Immense Journey.* Random House, New York. p. 207.

The evaporation of water from the leaves creates a lowered pressure within the leaves allowing the surrounding air pressure to force water and dissolved minerals up to the leaves from lower parts of the plant. Heat itself is also necessary to aid in the many diverse chemical reactions that occur in the maintenance of life, including photosynthesis. There are other processes involved also, but transpiration is mentioned at this point because it is an important result of work done by the energy received by a green plant, and furthermore, it is work done by energy that cannot be converted by photosynthesis into chemical energy.

The term energy, up to this point, has been used several times to convey a concept of its nature. Defined, it is simply the ability to do work. As in the green plant, the work may be moving water through the plant, or it may involve the building of complex organic structures. These organic substances, consequently, possess within their structures, a great deal more energy than the raw materials from which they were synthesized.

The conversion of radiant energy from the sun into chemical energy and heat illustrates a fundamental concept which unifies all of the sciences. This concept, called the *first law of thermodynamics,* states that energy may be neither created nor destroyed, but may be transformed from one type into another. Much of the activity of our own bodies is constantly engaged in converting some of the stored chemical energy into motion, electricity, and heat. Some organisms such as the firefly are even capable of converting some of their chemical energy into a special type of light.

Another fundamental principle describing energy relationships is the *second law of thermodynamics.* Briefly stated, this law relates that if left to itself, any system tends toward a state of greatest disorder. A high energy state means that a system possesses a high degree of order, a situation that is unstable unless a constant energy input is supplied to maintain it. If left to itself, it will tend toward a more disorganized form, and hence, a more stable form. A ball on an incline offers a simple analogy. If left alone, it will move from a position of higher energy to one of lower energy by itself. But if it is held there, it will remain at the higher energy level as long as the holding force is supplied.

Living systems possess a high degree of order. In fact, they are perhaps the most ordered systems known. The price for maintaining such a high degree of order is the expenditure of energy, obtained ultimately from the sun. In examining the second law of thermodynamics, we soon discover that as energy is passed along from one form to another, or from one chemical to another, less and less energy is available for use by an organism. Although the total amount of energy remains constant, as predicted in the first law, the amount that can actually perform work becomes smaller. The remainder is transformed into other forms such as heat, motion, electricity or possibly light as noted above. The usable energy, then, tends to run down unless there is a constant supply of energy from outside the system being received by the living organism.

Since much of the energy received by an organism is rendered unusable by the activities of that organism, we can easily see that it has less energy available to pass on to higher trophic levels. Hence, there is a practical limit to the number of links that a food chain can support. The unusable energy is dissipated into the environment. When the decomposers pass on the materials of life to again be used by green plants, the energy dissipation is virtually complete. Thus, the usable energy originally received from the sun, is eventually completely lost to the environment in a lower energy form, namely that of heat, and the prediction of the second law is achieved. Unlike the material aspects of a community, energy does not follow a cyclic path. Rather, it is a one-way street, with traffic on that street depending entirely upon a constant input from a single source.

ECOLOGICAL EFFICIENCIES

In 1951, E. I. Rabinowitch estimated that although local situations vary considerably, the plant world in natural conditions is able to utilize about two percent of the usable light energy which is actually absorbed. Remember that only about one-half of the radiant energy received is actually considered usable in the photosynthetic process. The efficiency, then, of the plant world is actually in the neighborhood of only one percent. Much of the energy is reflected from the plant surfaces or absorbed by the plant and converted into heat. Since approximately 70 percent of the oxygen of the atmosphere originates from photosynthesis occurring in the oceans, it is apparent that much of the usable solar radiation is absorbed by bodies of water. This energy is quickly converted to heat and is therefore lost for photosynthetic use. One estimate for the efficiency of photosynthesis in the ocean as a whole places the figure at 0.18 percent of the total radiant energy reaching the surface.

Once the usable radiant energy has been "fixed," that is, once it has been transformed into chemical energy, the transfer of usable energy from one trophic level to a higher one becomes considerably more efficient. It has been estimated by ecologists that on the average, organisms on one trophic level will pass on to the next higher level approximately ten percent of the energy which they receive from a lower level. However, this would still lead one to believe that living organisms possess a very low efficiency level, a belief that is contrary to popular ideas on the subject.

Although the efficiencies of man-made mechanical systems will vary considerably, a popular, yet reasonable figure that is often used for the efficiency of an automobile engine is 25 percent. It must be kept in mind, however, that this represents ideal efficiency during operation. No consideration is made for the energy input required for the construction, maintenance, and repair of the engine. Taking these energy requirements into consideration sharply reduces the actual efficiency of operation.

Living systems, on the other hand, are self-constructing, self-maintaining, and self-repairing. Energy input is constantly being used for these processes and efficiency levels are based on data which include these processes. Another consideration that must be made is that organisms, like engines, may sacrifice efficiency for speed. In other words, the evolutionary processes involved in natural systems may favor rapid growth for survival rather than high efficiency levels.

ECOLOGICAL PYRAMIDS

The most simple version of a natural system in which energy is passed from one trophic level to the next takes the form of a pyramid (Fig. 3-1). The important point illustrated in such a version is that each step of the pyramid is smaller than the one upon which it rests. Of the energy which an organism receives as food, normally less than one-half goes into the construction of new tissue. Some of this new tissue, such as bone, hair, and scales is not available to the next trophic level for energy since these are usually not digestible. Even some tissue that is available passes through the digestive system and is deposited in feces before it can be completely absorbed. Consequently, this too is lost as an energy source to that organism. Such material, however, is not entirely lost, for it then becomes available to the decomposer elements of the community.

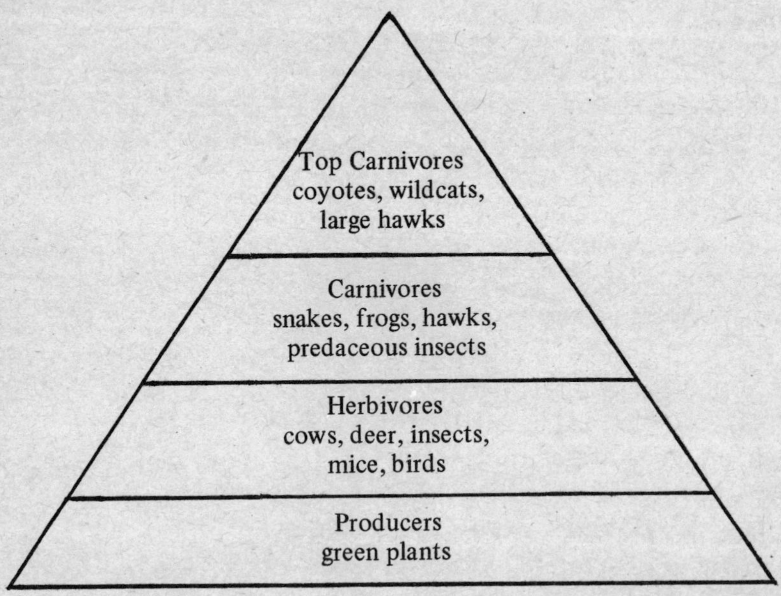

Fig. 3-1. A biotic pyramid. The upper levels must be supported by the lower levels.

The construction of a diagrammatic pyramid representing a natural community is usually based upon one of three criteria. A *pyramid of numbers,* such as that shown in Fig. 3-2A, illustrates a situation in which the actual numbers of organisms at each trophic level are depicted. Sometimes it is more convenient to weigh the organisms to get an estimate of the relative size of each level of the pyramid. In such a case, a *pyramid of biomass* is used (Fig. 3-2B). Biomass is defined as the total quantity of living material (usually expressed in dry weight) present at a particular time in a particular area.

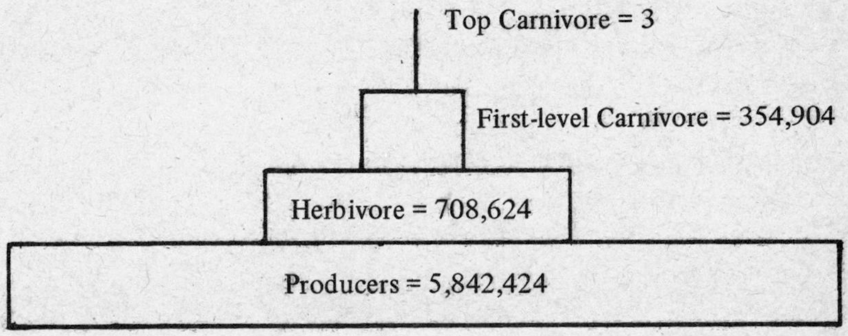

Fig. 3-2A. Pyramid of *numbers* in a one-acre grassland community. (From Hylander, 1956.)

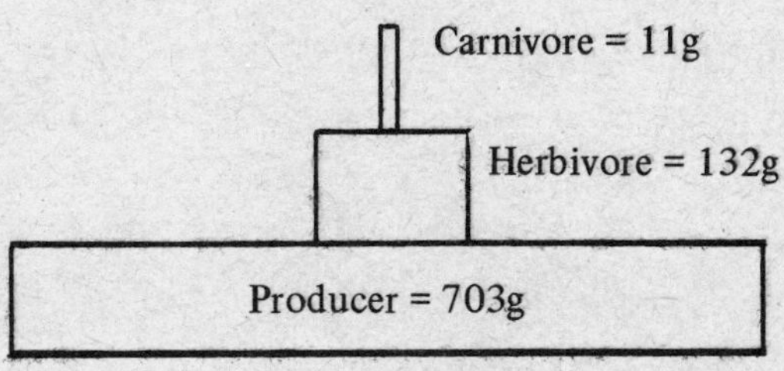

Fig. 3-2B. *Biomass* pyramid of a coral reef, Eniwetok Atoll. Numbers represent grams of dry biomass per square meter. (From Odum, 1955).

These two forms of ecological pyramids are similar in one important respect: the number or weight of organisms that can be supported at any given trophic level is dependent upon the *rate* at which food is being produced at the next lowest level. If food can be produced at a fast enough rate by a small number of organisms to supply the food requirements of a large number of organisms at the next higher trophic level, the pyramids may appear inverted. Such might be the case in which a single tree is the producer organism in a particular chain and a large number of insects feed upon the tree. In this case, the pyramid of numbers is inverted. The biomass pyramid may appear inverted in an aquarium situation in which a relatively small amount of algae is able to reproduce at a rate fast enough to provide food for a snail population.

The third type of pyramid, however, can never appear inverted. The *pyramid of energy* is based upon the amount of energy that is available at any given time. The transfer of energy from one organism to another involves a certain amount of loss as predicted by the second law of thermodynamics. Therefore, there must always be energy available at one trophic level in sufficient quantities to support the next level.

The number of organisms that can be supported in a given area for an indefinite period of time is known as the *carrying capacity*. It is based upon the available energy in that area, and in normal situations, is not exceeded except perhaps briefly. The concept of carrying capacity will be treated in more detail in Chapter 4 in connection with population dynamics.

In ecology, energy is usually measured in terms of the amount of heat that could be produced by completely burning the material and therefore extracting all of the usable energy. The *calorie* is the unit that describes the amount of heat needed to raise the temperature of one gram of water one Centigrade degree. In biological systems, this would usually be a rather large number, and so it is usually expressed in thousands of calories, or kilocalories (kcal). Thus, the amount of energy available in a particular area over a year's time might be expressed as kilocalories per square meter per year (kcal/m^2 /year). Figure 3-3 depicts an energy pyramid, and at the same time, illustrates the overall role of the decomposer element of the community.

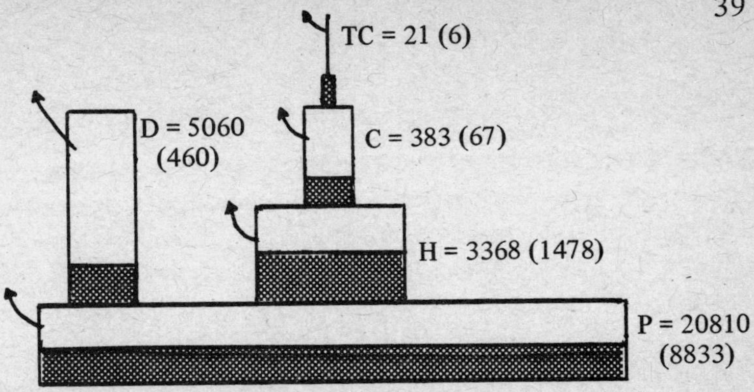

Fig. 3-3. Energy pyramid for Silver Springs, Florida, on an annual basis, in kcal/m^2/year. The energy which is actually fixed as organic biomass and which is potentially available as food for organisms of the next trophic level is indicated by figures in parentheses and the shaded portion of each step. Key to symbols: P, producers; H, herbivores; C, carnivores; TC, top carnivores; D, decomposers. (From Odum, 1953.)

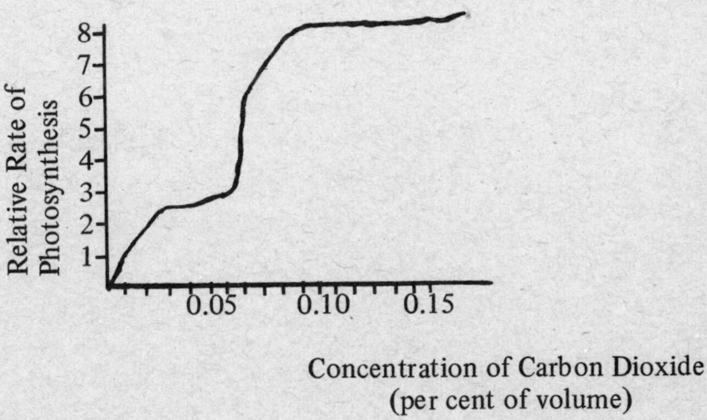

Fig. 3-4. As the concentration of carbon dioxide increases while other factors remain constant, the rate of photosynthesis increases until the concentration of carbon dioxide is no longer a limiting factor.

LIMITING FACTORS

As early as 1840, Justus Liebig clearly recognized the importance of minimal amounts of materials in the survival of plants and animals. His *Law of the Minimum,* as the concept came to be called, pointed out that the essential material available in quantities that most closely approached the minimum amount required for survival would be the factor that would limit the growth of that organism. Although carbon dioxide comprises only approximately 0.03 percent of the atmosphere, experiments have shown that when other factors (such as temperature, light, moisture, etc.) remain constant while the carbon dioxide content of the atmosphere is increased, the rate of photosynthesis by many green plants can be sharply increased. The atmospheric level of carbon dioxide would be considered a limiting factor in the growth of these plants (Fig. 3-4). The curve quickly levels off after a concentration of 0.08 percent carbon dioxide is reached, indicating that further increases in the rate of photosynthesis are limited by another factor. Figure 3-5 confirms this hypothesis, showing several curves resulting from increasing the carbon dioxide concentration at various light intensities.

Although Liebig's work was originally concerned with the chemical materials required by organisms to survive, biologists were quick to recognize the importance of the concept with respect to other environmental factors such as temperature, humidity, light intensity, light quality, and others. Then, in 1913, V. E. Shelford further expanded the concept to include the maximal limits of environmental factors. Not only would factors approaching minimal levels limit growth, he reasoned, but many factors might also be limiting if supplied in amounts surpassing a critical maximum. Thus, a range of tolerance, lying between the minimal and maximal levels of survival for a given factor can be distinguished. As long as conditions fall within the limits of tolerance, the organism will survive and reproduce. This concept became known as Shelford's *Law of Tolerance.*

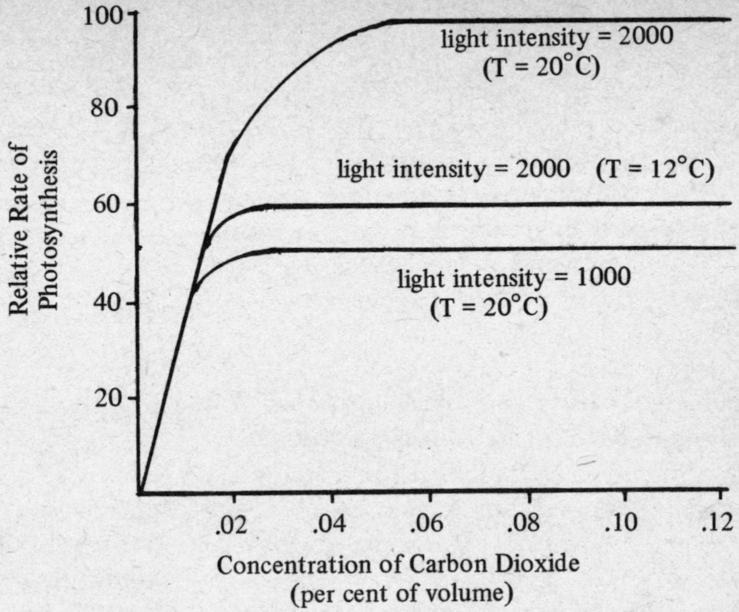

Fig. 3-5. Photosynthesis can be further increased after carbon dioxide no longer is a limiting factor by increasing the light intensity. Temperature can also be a limiting factor as shown by the middle line. (From Rabinowitch, 1951.)

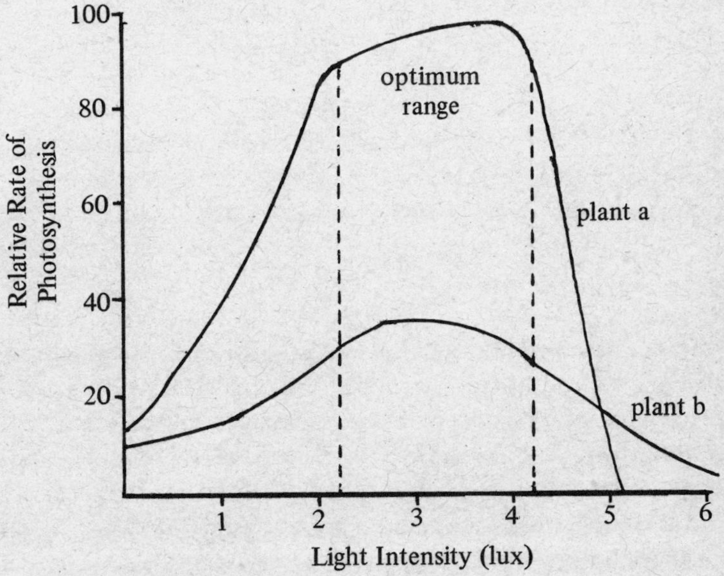

Fig. 3-6. Two different plants may possess different ranges of tolerance to light intensity. Even so, the optimum range for both plants is roughly the same. Which plant is more tolerant to sunlight?

Interestingly enough, within the range of tolerance that an organism seems to possess for a certain environmental factor, there is an optimum range in which maximum growth, reproduction or activity is obtained. Figure 3-6 shows the range of tolerance to light intensity possessed by two hypothetical green plants. Within their respective ranges, there is a fairly well defined area which indicates the *optimum range* for these plants; that is, the range at which maximum photosynthesis occurs. Note that the two plants do not possess the same range of tolerance to light intensity, but that the optimal range is approximately the same for both.

Limiting factors applied to one trophic level may well exert their effects on other levels indirectly. A reduction in the numbers or kinds of plants due to some limited factor in the environment will result in less energy being made available to higher trophic levels. On the other hand, abnormal increases in a consumer population (an insect, for example), while temporarily giving a broader base to a carnivore trophic level above, may eventually destroy the producer level sufficiently to upset the entire pyramid for a long period of time.

The study of various conditions as limiting factors can be a rather exhausting experience for a variety of reasons. It often develops that limiting conditions defined in the laboratory are not limiting in natural situations. Apparently, the interaction of environmental factors provides the organism with a different range of tolerance. Consequently, the ecologist must be careful about applying laboratory data to a natural situation.

Equally hazardous, however, can be the practice of applying information gathered in one area to the same species in another area. Some organisms are distributed over a large area. The distribution of the leopard frog, for example, ranges from Canada deep into Central America. Environmental conditions at one end of the range can be quite different than those at the other end. Although the frogs are of the same species, local populations will have minor variations which enable them to be well-adapted to the particular set of environmental conditions in their area. Consequently, each population may possess a range of tolerance for a given condition that is slightly different from that of an adjacent population. Furthermore, differences

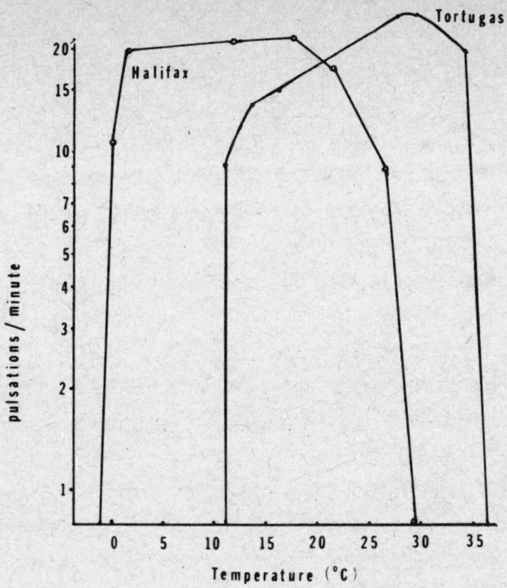

Fig. 3-7. The relation of temperature to swimming movement in northern (Halifax) and southern (Tortugas) individuals of the same species of jellyfish. The natural habitat temperatures were 14° and 29°C, respectively. Note that each population is acclimated to swim at a maximum rate at the temperature of its local environment. (From Odum, 1953, after Mayer.)

Fig. 3-8. Diagrams of relative ear length in the jackrabbit. A, Texas; B, South Dakota; C, Canada.

between populations at the extremes of the range of distribution may be quite significant (Fig. 3-7). Locally adapted populations are usually referred to as *ecotypes*; that is, they are populations within a single species that have acquired slight variations enabling them to be better adapted to a particular habitat. Local variations may appear in the size of the extremities (ears, legs, tails) of warm-blooded animals. Ecotypes of the jack rabbit, for example, in southern reaches of the range of distribution usually exhibit longer ears and legs than members of the same species living in more northern portions of the range (Fig. 3-8). This phenomenon is generally regarded by ecologists as *Allen's Rule.* Smaller extremities help to conserve body warmth for animals living in cold climates, while large extremities on animals living in warmer areas of the world are advantageous in that they help to remove excess heat from the body.

Another principle, called *Bergmann's Rule* states that among warm-blooded animals, the body size increases poleward from the equator. This increase in size reduces the surface area to volume ratio of the body and thus tends to retard heat loss in colder climates. Both of these principles become obvious when one considers the animals of the extreme polar regions. Polar bears, seals, and walruses possess relatively massive bodies with reduced appendages such as legs, ears, and the like. These features must be regarded as physical adaptations to life in conditions where heat conservation is a matter of life and death.

Other corollaries to Shelford's law have also been described and documented by ecologists. Some organisms will have a wide range of tolerance for one environmental factor and a relatively narrow range for another. Certain freshwater fish may possess a wide range for temperature but can only tolerate a rather narrow range of salt concentration. Furthermore, research has shown that if conditions approach the tolerance limits for one factor, the range of tolerance for other factors may be sharply reduced. You are undoubtedly well aware that you are able to tolerate a wider temperature range if the air is dry. For example, when the temperature is particularly high, and the humidity is also high, evaporation of moisture from the body is retarded, and the cooling mechanism is hampered.

If conditions continue for a very long time, survival may become a very real problem. Local weather bureaus often give a comfort index with their reports which reflect the combined effects of temperature and humidity. Combined wind and temperature effects similarly reduce the tolerance to cold temperatures, and are given in weather reports as the chill factor.

Another corollary reveals that organisms with wide ranges of tolerance for all factors are likely to be most widely distributed. This is only reasonable since environmental conditions outside a particular habitat may offer too much variation to a narrowly-adapted organism, a situation which such an organism could not survive.

One final point will be made here concerning the optimum range. It is not uncommon that an organism is not actually living within what has been determined to be the optimum range for its survival. Careful study of the complete picture may reveal that other factors of the environment are more important, and the interaction of these factors will dictate the actual conditions under which the organism will be found. The factors which may affect survival may be either abiotic or biotic in nature, and it is often easy to overlook possible facets of interaction. Perhaps one of the most common types of examples of a situation of this nature would be plants whose optimum range of light intensity is relatively high. However, these plants are often observed growing in more shaded areas because their tolerance to the heating effects of direct sunlight is too low.

QUESTIONS FOR DISCUSSION

1. In terms of energy relationships, how is radiant energy involved in the water cycle?
2. Suggest some everyday examples that would illustrate the first and second laws of thermodynamics.
3. Explain the concept of an energy pyramid.
4. List three types of energy pyramids. Describe, in general the conditions under which two of these types of pyramids might appear inverted. Why can the third type never be inverted in a well-balanced community?

5. Describe the basic difference between the flow of energy through an ecosystem and the flow of materials.
6. Explain briefly the meaning of Liebig's Law of the Minimum.
7. How did Shelford expand the concept of Liebig's law?
8. What is meant by the term optimum range? How does this compare with the range of tolerance?
9. What is an ecotype? Why can ecotypes be regarded as members of the same species in spite of variations between populations?
10. Why must the range of tolerance for other factors be considered when the critical extremes for one factor are approached?
11. Why is it reasonable that organisms with wide ranges of tolerance for many factors be relatively widely distributed?
12. Why is it relatively common to discover that many organisms are living under conditions that are not within the optimum range for survival?

INVESTIGATIONS AND ACTIVITIES

1. Design and carry out an experiment by which the photosynthetic efficiency of an alga population such as *Chlorella* can be altered. Consider such problems as (a) what means can be used to evaluate production? (b) what means for determining efficiency can be used? (c) how does light intensity alter efficiency? (d) how does culture size affect efficiency? (e) what problems are encountered with respect to contamination? (cf. Tamiya, 1957)
2. Constant mention of "usable light energy" is made in the chapter. Prepare a report that relates in more detail what is meant by this phrase. Attempt to describe its meaning in terms of the nature of light. Can you design an experiment to demonstrate that some parts of visible light are more usable than others?
3. Some attention has been focused upon the indirect use of heat in the photosynthetic process. Design and carry out an experiment to show that the rate of photosynthesis is dependent upon heat. Consider such problems in your experiment as (a) what criterion can be used to indicate

that photosynthesis is being carried out and (b) what is meant by rate?

5. Look up the term carrying capacity as it is used in an ecological sense. Describe its meaning using examples to illustrate your description. In a brief report, explain why the carrying capacity of an area is less for higher trophic levels of an energy pyramid than that for lower levels.

RECOMMENDED READING

Bassham, J. A. 1962. *The Path of Carbon in Photosynthesis.* Sci. Amer. 206(6): 89-100.

Bonner, James. 1962. *The Upper Limit of Crop Yield.* Science. 137(3523): 11-15.

Gaffron, Hans. 1965. *Photosynthesis.* BSCS Pamphlet Series, No. 24. D.C. Heath and Co., New York.

Knight, Clifford. 1965. *Basic Concepts of Ecology.* MacMillan Co., New York.

Franck, James, and Walter E. Loomis. 1949. *Photosynthesis in Plants.* Iowa State College Press, Ames.

Odum, Eugene P. 1953. *Fundamentals of Ecology,* 2nd ed. Saunders Co., Philadelphia.

Rabinowitch, Eugene I. 1951. *Photosynthesis and Related Processes,* vol. 2, part 1. Interscience Publishers, Inc., New York.

Simpson, G. G., and W. S. Beck. 1965. *Life: An Introduction to Biology.* Harcourt, Brace and World, Inc., New York.

Stephens, William M. 1968. *The Turtle Grass Community.* Natur. Hist. 77(2): 50-57.

Tamiya, H. 1957. *Mass Culture of Algae.* Annu. Rev. Plant Physiol. 8:309.

Wald, George. 1959. *Life and Light.* Sci. Amer. 201(4): 92-108.

Weiz, Paul B. 1963. *The Science of Biology,* 2nd ed. McGraw-Hill Co., Inc. New York.

Woodwell, George M. 1967. *Toxic Substances and Ecological Cycles.* Sci. Amer. 216(3): 24-31.

4

"Being a specialized entity, every organism depends on other organisms for some essential product or process; no organism can survive in strict isolation."

Weisz*

Population Ecology

Although the individual organism may be the fundamental unit of an ecosystem, the ecology of a single individual cannot be adequately studied without relating it to other members of his species and other species living in the community. Consequently, ecologists usually use the population as the fundamental unit for study purposes.

Populations are unique in the spectrum of ecologic levels because they possess certain descriptive characteristics that no other level possesses. Such characteristics are descriptive of specific groups and not of individuals. Individuals cannot be described in terms of their density, distribution, natality (birth rate), mortality (death rate) or cyclic fluctuations in numbers. These are characteristics of groups of organisms which are homogenous; that is, these attributes are used to describe the characteristics of a population which is comprised of a single type of organism. At higher levels of organization, the community, for example, there are many types of organisms, and the attributes describing populations are no longer appropriate.

POPULATION DENSITY

The *density* of a population is the total number of individuals living in a particular area at a specific time. Such a definition does not really separate the concept of the population, per se,

*Weisz, Paul B. 1963. *The Science of Biology.* McGraw-Hill Book Co., Inc., New York. p. 119.

from that of density. In discussing the concept of density, we usually reduce the area to a fundamental unit appropriate for the organism we are discussing. For example, if we are discussing the population of the United States in a particular year, the area is understood and is a general area rather than one of a unit nature. On the other hand, if we were to discuss the density of the population of the United States, we would not be talking about the total population, but rather about the number of people occupying a unit of area, say a square mile. Whereas the population would be 200 millions of people, the density of the population would be 53 individuals per square mile (53 individuals/mi^2). This means, simply, that if the population were to be spread evenly over the existing surface of the country, there would be 53 people within each square mile of land. To calculate the density of any given population is a relatively simple task. The word "per" means divided by: thus, the number of individuals is divided by the total area in which the individuals are found. The formula $D = \frac{N}{S}$ describes the situation mathematically, where the population density, D, is shown to be equal to the number of individuals, N, divided by S, the area being occupied by those individuals.

Knowledge of the density of a population may give us some valuable information regarding a population. However, unless some information regarding the *distribution* of that population is also given, we might have a very lop-sided view of that community. We have seen that knowing the population and area of the United States at a given time, the density could be calculated. But the population is not so evenly distributed throughout the country, and use of density alone could give a rather false impression of the country to an uninformed person.

The same situation might occur while studying a natural community. Suppose that we sampled the population of an area by dividing it into ten equal plots (quadrats). Suppose, further, that we determined that thirty plants made up the total population of a certain shrub species in that area. The density, as determined by the formula would be $\frac{30 \text{ individuals}}{10 \text{ quadrats}}$ or 3.0 individuals per quadrat, and a mental picture of the community might appear as in Figure 4-1. However, if all of the

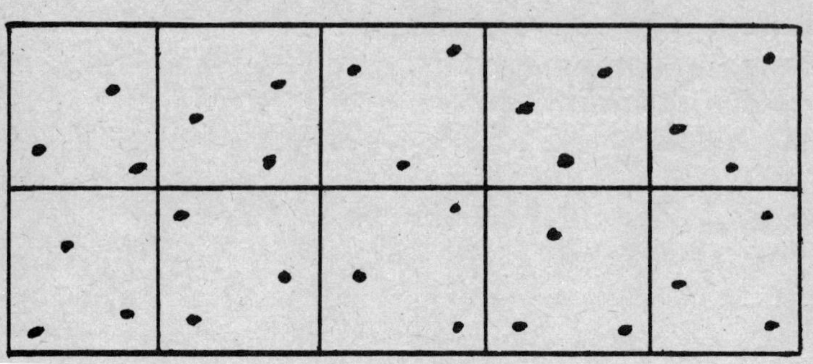

Fig. 4-1. Diagram of an area divided into ten quadrats with individuals scattered evenly throughout the area. Density of the population is three individuals per quadrat.

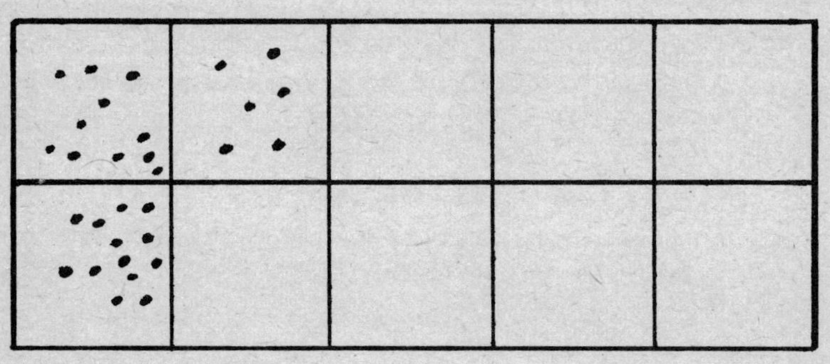

Fig. 4-2. If the population shown in Fig. 4-1 had been distributed as in this diagram, the density of the population remains at three individuals per quadrat. However, an entirely different impression regarding the population is obtained.

Monoculture, while highly efficient for producing food for large numbers of people, is an open invitation for high populations of pests.

shrubs had actually been found in only three quadrats (Fig. 4-2), then certainly a different picture of the community is obtained.

By now, it should be quite apparent that density and distribution of individuals are especially important in determining the amount of interaction that will occur between individuals of the same or other species. When densities are high and distribution of organisms is widespread for a single species, overpopulation becomes a threat for several reasons. Competition for food or nutrients becomes a factor for survival. Similarly, competition for shelter and mates is heightened. Predators become attracted to the area and their populations tend to increase because of the availability of food. Parasites may also increase and thereby reduce the resistance of the population in question to environmental factors as well as to predators and disease. The population will, under natural conditions, be sharply reduced by a combination of such factors.

This is one of the reasons why farmers in the United States are constantly engaged in a struggle for survival against the elements of natural ecological processes. Vast areas of farmland are devoted to only a few crops. Examples of this are easily understood when you visualize the concepts of the wheat belt of the Dakotas, Nebraska, Kansas, Oklahoma, and Texas, and the corn belt of Iowa, Nebraska, Kansas, and northern Missouri. Such large concentrations of a relatively few crops are constant invitations to predators, plant diseases and parasites. To fight the natural influx of such pests, the farmer must arm himself with pesticides, constantly fertilize the depleted ground, and be on the lookout for new crop varieties that possess special resistances to the invading armies of diseases and pests.

When the population of a particular organism is lower than that which can be supported in a given area, the result is often an increase in food, shelter, and eventually an increase in the population itself. On the other hand, underpopulation may be as disastrous as overpopulation. Some organisms apparently depend upon group activity for their very survival. During winter months, the bobwhite quail maintains group activity in the form of a covey. Birds cluster together for warmth with the outer members slowly being worked to the inside of the covey. When the population is reduced to a critical minimum, due to over-hunting, disease, high populations of predators or other factors, the entire covey may perish. This same behavior may also be observed among cattle during particularly cold periods.

The musk ox, once nearing extinction in the wastelands of the far north, depends upon the defense activity of the herd to protect itself from its primary predator, the wolf. When alarmed, the bulls form a circle, facing out, with the cows and calves on the inside. In this position they present a wall difficult for a wolf to penetrate. When a herd reaches its critical minimum, the defense mechanism fails, and the herd is vulnerable to decimation. Such a defense, of course, renders the musk ox extremely susceptible to slaughter by men, and by 1917, the animals had become so rare that the Canadian government ordered their protection. Other governments have since followed suit, some as late as 1951. From an estimated population of 500 at that time, the Canadian population has since approximately tripled with similar results in other Arctic regions. Today, the musk ox is slowly beginning to spread back into areas where it previously existed.

But such fortune did not favor the relatively recent demise of two species of birds. The last heath hen died on Martha's Vineyard off the coast of Massachusetts, in 1931, after attempts to reestablish a population failed. The heath hen, by 1890 had been reduced to a paltry 200 birds and extinction seemed imminent. After coming under protection, their numbers increased to some 2,000 twenty-six years later, only to have a severe fire destroy possible nesting and shelter sites as well as food. This factor, coupled with adverse weather conditions during 1916, reduced the population to 50 breeding pairs. Further attempts to establish a flourishing population failed, and the colony was reduced to a single male by 1928. Apparently, the critical minimum had been reached, and although there seemed to be a sporadic resurgence of the bird, the ultimate end was in sight by the time of their release on the island.

The fate of the famed passenger pigeon was equally disastrous, and also at the hand of man. Many writers as recent as 1875 recount instances where the sky would be literally blackened by the flocks of passenger pigeons. Because their great numbers presented serious problems to man, they were wantonly slaughtered as pests as well as for food. Again, the inevitable end was in sight by the time the realization that wholesale decimation had gone too far, and the last surviving specimen finally died in a Cincinnati zoo in 1914. Although most biologists will agree that several factors, including environmental alterations caused by the removal of large expanses of forest as well as reduction of the population below the critical minimum were factors in the extinction of the passenger pigeon, there is little doubt that underpopulation played a definite part.

POPULATION GROWTH AND CONTROL

Several factors contribute to the dynamics of population growth. Basic to its growth, of course, is the organism's inherent ability to reproduce, or *natality*. If the natality is expressed in terms of new individuals over a unit period of time, then this number becomes known as the *natality rate*. Although every population has a maximum natality rate, it is seldom, if ever, actually achieved. Against any population's efforts to increase, the environment is constantly offering a resistance in the form of predators, disease, unfavorable weather conditions and lack of food, shelter or mates. In spite of such resistance, populations will grow and continue to do so until one of the resisting forces is sufficient to alter the natality or cause the *mortality* (number of deaths) to exceed the natality.

For convenience, particularly in human population studies (demography), the natality rate is often expressed as a rate per unit of population. For example, if we were to state that the natality rate in the United States was 3,533,000 individuals per year, we would only have a vague basis for comparison when we learn that the natality rate for India is some 12,778,000 individuals per year. Consequently, the natality rate is usually expressed in terms of rate per 1,000 population. Compared on this basis, we see that a comparison of the two countries based on equal populations, there is in fact a greater natality rate in India. The United States would show a natality rate of 17.9 individuals per year per 1,000 population against approximately 42 individuals per year per 1,000 population in India.

Growth curves are often employed by ecologists to illustrate the features of population growth. Figure 4-3 depicts an ideal curve that would result as a population grows from a relatively few individuals to a size where space and food become limiting factors. Natural populations will then cease to grow in size as shown in the graph.

The reduction of the relatively steep slope of the curve as the population becomes limited may be due to any one or a combination of factors. As food becomes scarce, mortality of individuals may increase, slowly at first, but at a greater rate as the population continues to grow. When the mortality rate reaches a point where it equals that of natality, the population no longer expands. Often, the situation does not neatly stabilize at this point, and the mortality rate continues to increase, being reflected on the graph as a sharp drop in population (Fig. 4-4).

Lack of food is not the only factor which may cause a sharp rise in mortality. The rapid increases of wastes from respiration and other body processes may exert a profound influence upon a population. A growth curve for a population of bacteria may show a death phase which reflects both lack of food and the accumulation of toxic waste materials. Increased environmental resistance in the form of disease, parasites, and predators may also be important factors which eventually result in a decline in the growth curve.

Growth of natural populations can also be affected by other factors. Thus far, we have assumed that the populations studied are closed; that is, their growth curves reflect only the effects of natality and mortality. Populations in nature, however, are often subject to the influences of immigration and emigration. New individuals may immigrate to an area with a smaller population density where conditions for survival are more suitable. Similarly, under conditions of overcrowding, decreased food, or other factors detrimental to survival, individuals may leave or emigrate from the population.

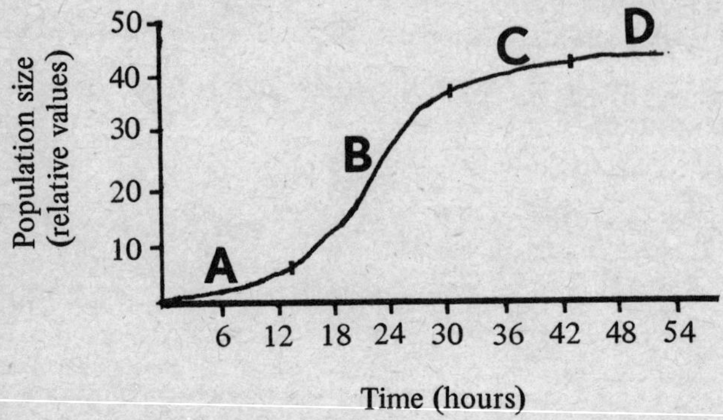

Fig. 4-3. A typical population growth curve showing basic phases of growth. A represents the lag or acceleration phase; B, the logarithmic phase; C, the negative acceleration phase; and D, the maximum population.

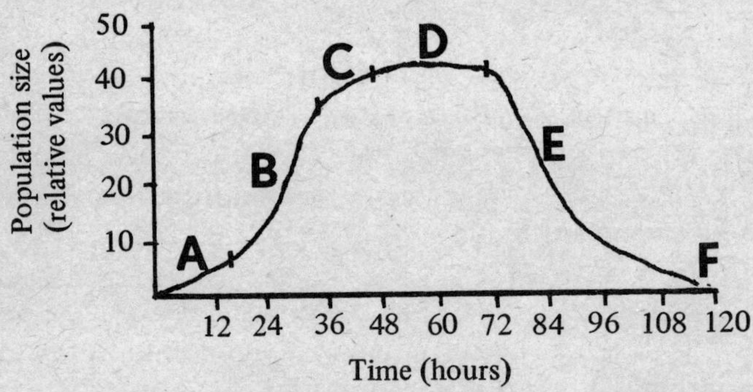

Fig. 4-4. Typical growth curve that might represent a bacterial population. A-D are features which are described in Fig. 4-3. E represents the death phase, and F is the point of extinction.

Man may also exert his influence upon the control of the size of naturally-occurring populations, directly, as in the case of the heath hen, passenger pigeon and musk ox, or indirectly as in a classical case involving deer in the Kaibab National Forest in Arizona. In an effort to increase the deer population, the Forest Service initiated a massive predator control program in 1907. Neighboring ranches had been noting losses in their sheep and cattle herds, indicating to the Forest Service that predator control was warranted. Although the area held a herd estimated to be 4,000, the carrying capacity was estimated to be closer to 30,000.

In only three years, the success of the program was quite evident. The deer population had climbed to 30,000 and was not showing signs of stabilizing. In sixteen years, 674 cougars, 11 wolves and over 3,000 coyotes had been killed under the program and the deer population had exploded to over 100,000 by 1924.

But population growth could not continue at such a rate, and despite the continued optimism of the Forest Service, ecologists warned of impending disaster. The winters of 1924-25 and 1925-26 came early and proved to be extremely severe. Forage for the deer became scarce, and evidences of widespread famine among the overexpanded population became apparent. The earlier predictions of the ecologists began to come true as the winters progressed. By the spring of 1926, over 60,000 deer had starved to death! The experiment in game management had been turned into the predicted disaster. Food had not been a limiting factor to the deer population before the predator control program. But without the checks on population growth supplied by predators, the number of deer exceeded that which the range could support. Characteristic of a situation such as this, when the food reserves began to be depleted, the animals will suffer lowered resistance, but not immediate death. There is a resulting lag in time between the depletion of food and the onset of mass death by starvation.

Some organisms are capable of controlling their populations themselves. When the food supplies approach a level at which they may become a limiting factor, natality may be reduced, and the population ceases to grow. Although the food factor is important, it is important only indirectly since starvation is not present. V. C. Wynne-Edwards, a noted Scottish ecologist at the University of Aberdeen, has summarized the situation:

> . . . .Thus, the threat of starvation tomorrow, not hunger itself today, seems to be the factor that decides what the density of a population ought to be.

Studies on the ovaries of mature, but non-breeding fulmars reflect sexual inactivity and indicate that population pressures may prevent normal production of eggs, resulting in population control.

Specialized behavior patterns represent important means by which self-control of population is attained. Behavior is a very complicated device used by many animals to regulate a large segment of their daily activities. An enormous variety of behavior patterns have been described for many animals. Such patterns govern virtually every aspect of an animal's existence from feeding to mating and the raising of offspring. In this context, its importance as a population control mechanism should not be underestimated.

Population pressures may result in certain members of the group immigrating to new areas where pressures are less severe. Some animals may travel in herds with a single male leader. Males born into the herd, after reaching maturity, are driven away by the dominant male. Horses are perhaps the most common examples of a situation of this type. Other examples of pressures to emigrate may be less well known, but the effect is the same. Dispersal of individuals into less populated areas helps to ensure an adequate food supply. Among the birds, dramatic behavioral activities during the breeding season may act as a powerful stimulus to force less aggressive males to emigrate.

In some organisms, when emigration is not possible, and reduced natality is not sufficient to reduce population growth where food is limited, other, more stringent measures may be employed to control population. Among mice and rats, it is not uncommon for cannibalism to occur when the population begins to reach the critical maximum. Some investigators have found that in spite of an abundant supply of food, when emigration is prevented, overcrowding itself may become a factor resulting in cannibalism.

Many animals possess innate breeding patterns in which only selected individuals may participate in breeding activity. In many species of birds such as the gannet and the fulmar mentioned above, a reduced food supply results in an automatic reduction of breeding activity. The less aggressive individuals act as a reserve and replace losses in the breeding colony when and if they occur.

Another outstanding example of this type of limited breeding activity is described by Robert Ardrey in his book, *The Territorial Imperative.* The Uganda kob, an animal whose American counterpart might be the pronghorn, exhibits a very carefully ritualized mating behavior in which only twelve to fifteen males of a herd numbering over one hundred may take part. Within the range of a given population, the herd maintains a breeding ground where all breeding for that population takes place. High priority males have each established a small circle, approximately fifty feet in diameter, within the confines of the breeding ground, which they actively defend against other males of the herd. The remaining males have free range throughout the feeding grounds, but are not capable of participating in any breeding activity unless they are able to eventually secure a spot on the breeding ground. A low priority male may win a spot by successfully challenging one of the high priority males, or he may step into a vacated spot when one of the dominant males leaves to feed. However, the new occupant must answer to the spot's "owner" when he returns.

That territoriality is a factor in population control among animals has been known for some time. Especially noted in birds, breeding pairs will stake out a territory which they guard by any number of a large repertoire of means, including special territorial songs, elaborate display behavior or bright plumage. Seldom does actual fighting occur. Can you think of possible advantages to the species of such complex behavior patterns as an alternative to fighting?

Many plants also possess mechanisms which prevent overcrowding where water or nutrients may be in limited supply. The creosote bush, a common plant of the arid regions of

Plants often secrete chemical substances into the soil around them that prevent the germination of other plants in close proximity to them. This adaptation as seen in the creosote bush helps to ensure that there is adequate moisture for the existing plant, and results in an evenly spaced community.

North America, apparently secretes a chemical through its roots that inhibits new plants in the immediate vicinity. Seeds may germinate and produce new plants only when the distance from existing plants places the seedling outside the area which the inhibitor is able to penetrate. Even a superficial observation of a desert covered by creosote bush will demonstrate the remarkably evenly-spaced plants.

POPULATION FLUCTUATIONS

The ecologist is especially interested in the changes in population density that might occur over a period of time. Quite often, the density and distribution are not so important in themselves, and the ecologist must be aware of changes in density which serve as indicators of the state of health of a community. By graphing the density of a population over a period of time, he can keep abreast of population tendencies within a study area. Such a graph (Fig. 4-5) might share certain characteristics with that of a population graph, but it is important because it can be used to help determine a theoretically optimum number of individuals which can be supported in a given amount of area.

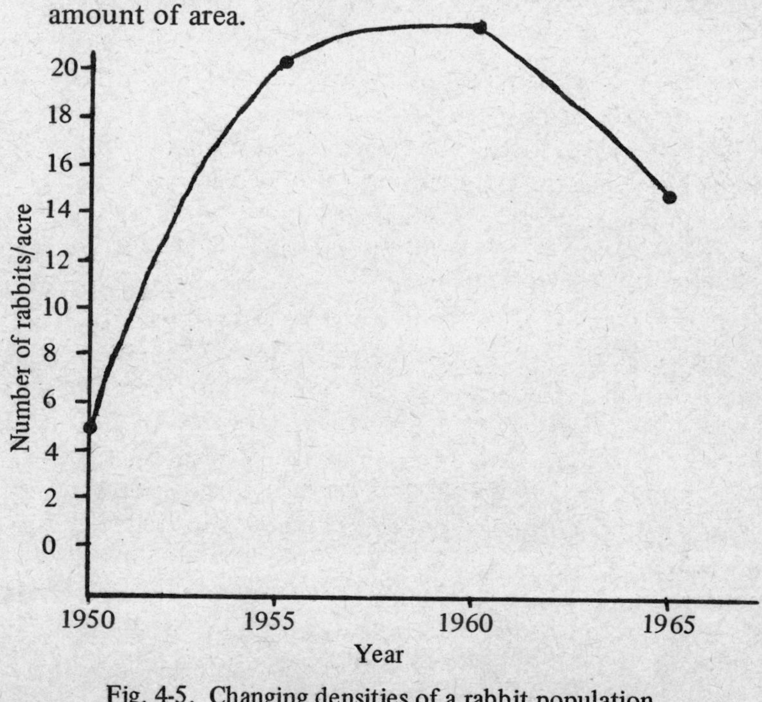

Fig. 4-5. Changing densities of a rabbit population.

Suppose that a sampling of rabbits in 1950 revealed their density to be five individuals per acre. A second sample in 1955 showed that the density had increased to twenty rabbits per acre. Subsequent samples in 1960 and 1965 indicated densities of 22 and 15 rabbits per acre, respectively. Figure 4-5 shows this data plotted graphically. Note that there was a rapid population growth in the five-year period from 1950-55, but a less rapid growth in the next five years. From 1960 to 1965, the population showed a marked decline. Since we are discussing a change of a characteristic over a period of time, we are really discussing a *rate*. In this case, we are discussing a rate of change in the density of rabbits; that is, we are interested in knowing the amount of change in density per year. Thus, we would describe the rate of density change in the period 1950-55 as being 15 rabbits per acre per five years. More simply, this would be stated as three rabbits per acre per year. Mathematically, we can describe this calculation as follows: the change in density, ΔD (read as "delta D"), during the five-year interval is divided by the change in time, ΔT. Hence,

$$\text{Rate} = \frac{\Delta D}{\Delta T} = \frac{15 \text{ rabbits/acre}}{5 \text{ years}} = 3 \text{ rabbits/acrc/year.}$$

The next five-year interval indicates that the rate of density-change is two rabbits per acre per five years, or 0.4 rabbit/acre/year. Since the value is still positive, the density is shown to be increasing but not at such a rapid rate as observed in the initial five-year period. This situation is verified by the graph. What is the rate of density decrease for the next five years? Is your answer verified by the graph?

Not all populations will exhibit a simple growth pattern. As we have seen, once a critical population level is attained, a variety of factors may combine to prevent it from continuing its growth. Natality and immigration may be offset by mortality and emigration, resulting in a relatively stable population. But even in organisms which do exhibit such population stability over a long period of time, there will be fluctuations in the population density.

One of the most noted examples of fluctuations in natural populations is that of the lynx. Pelt records kept by the Hudson

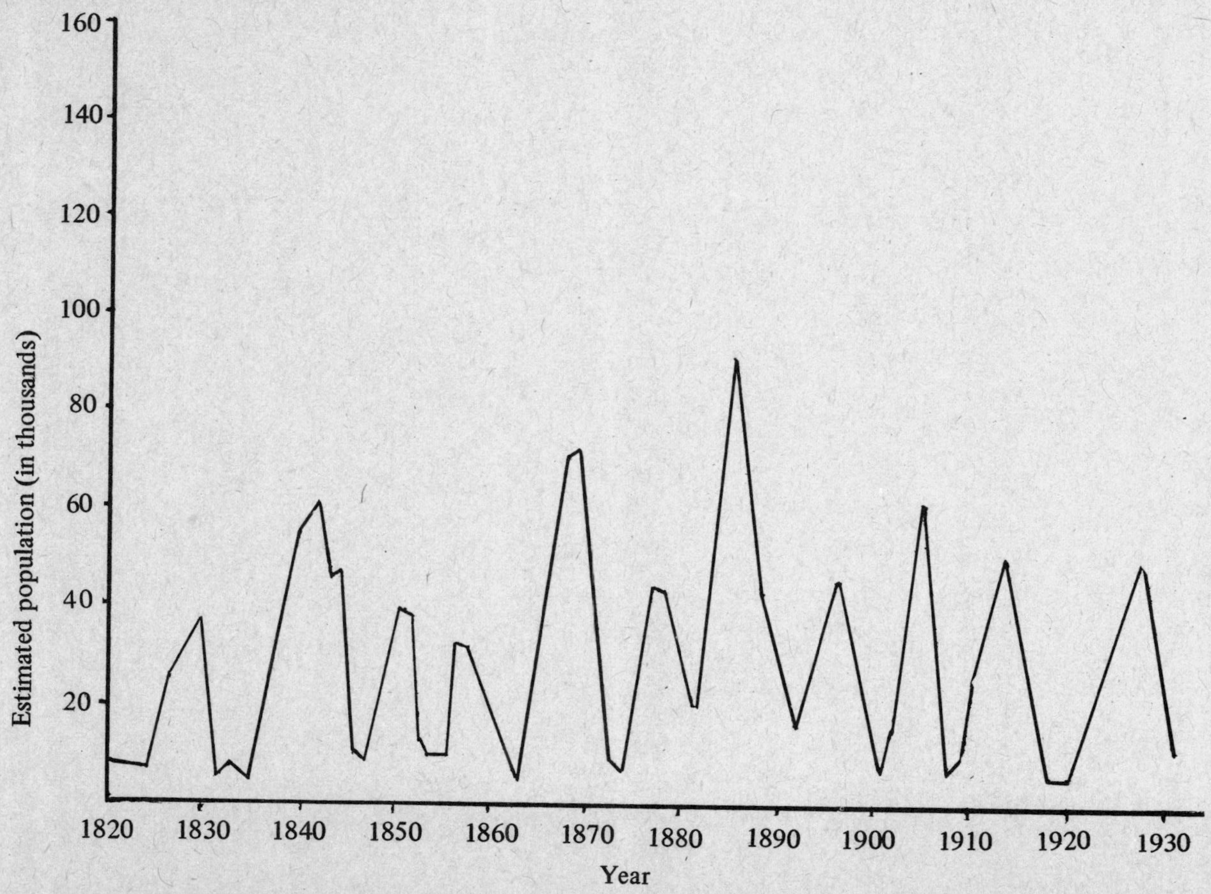

Fig. 4-6. Fluctuations in the population of lynx as reflected by pelts reported collected by the Hudson Bay Company of Canada.

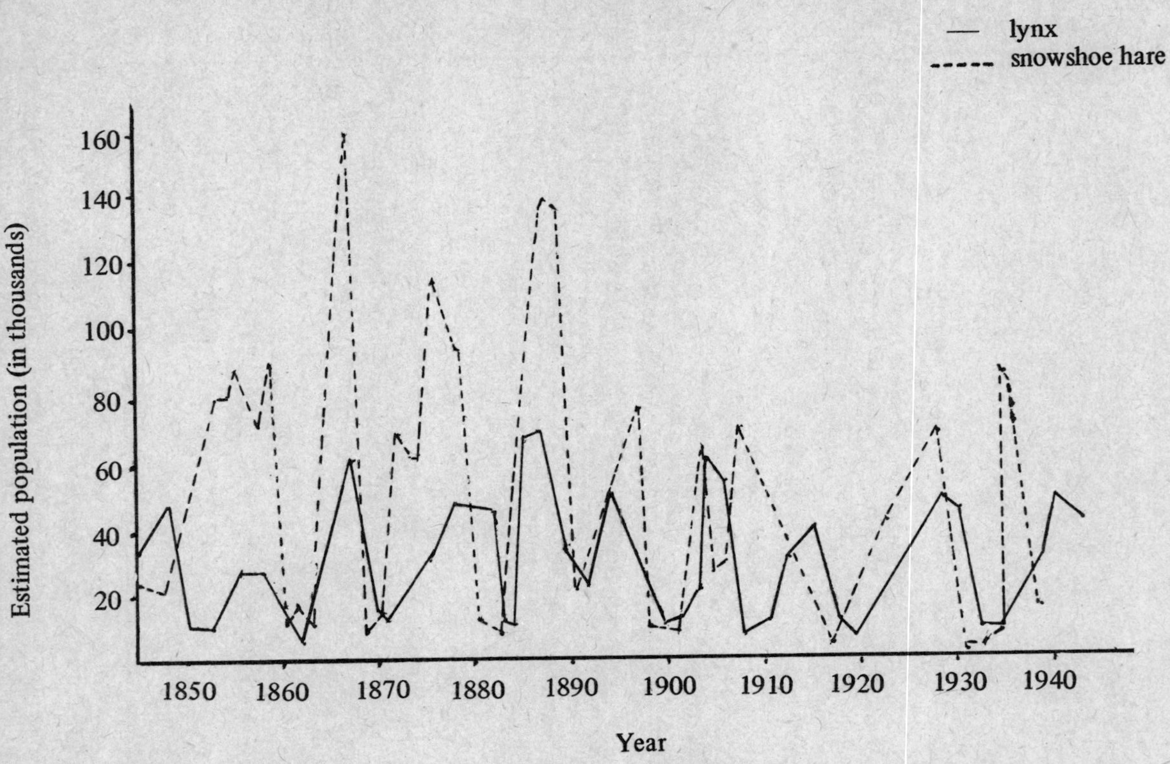

Fig. 4-7. A comparison of the populations of lynx and the snowshoe hare as reflected by pelts reported collected by the Hudson Bay Company of Canada.

Bay Company of Canada since 1800 show that the lynx reaches a population peak every 9-10 years (Fig. 4-6). Immediately following such peaks, periods of rapid decline in lynx pelts reflect a decreasing population during which time the lynx become rather rare.

In an attempt to determine the factor or factors responsible for such fluctuations, the ecologist must be aware of many possibilities, and consider some of the following questions: Are there climatic conditions which vary with regularity which could affect the lynx? Is there evidence for regular alterations in the diseases which affect the lynx? What fluctuations might occur in the food supply?

To answer these and other related questions, the ecologist must make a carefully detailed study of the natural history of the lynx. In so doing, he discovers, among other things, that the principal food source is the snowshoe hare. A review of the population fluctuations of the snowshoe hare since 1800 reveals a pattern strikingly similar to that of the lynx (Fig. 4-7).

Note that although the two curves fit together fairly well, they do not coincide exactly. The snowshoe hare reaches a peak in population a year or two before that of the lynx. The low point for the hare also occurs just before that of the lynx. Apparently there is some relationship here with food being a controlling factor in the lynx population. The predator-prey relationship, in turn, controls the hare population. As the hare population increases, it provides a larger food supply for the lynx. With an increased food supply, the lynx population can also increase. What effect can the increased lynx population have on the hare population? Are there any factors other than a large lynx population that might result in a sharp decline in the hare population?

Some questions might be raised as to why the peak in the hare population just precedes that of the lynx. Here, again, a thorough knowledge of the natural history of the organisms involved is necessary to determine a satisfactory answer. If the population of the lynx is controlled directly by the available food supply, then high numbers of hares would result in greater breeding activity in the lynx population. When the hare population begins to decline, due to increased predation as well as disease, large numbers of lynx would still be in embryonic stages of development. The newborn lynx would add to the population after the peak of the hare population and their numbers would not be recorded in pelt records until the following year.

It is important to realize that demonstrating a remarkable similarity and apparent relationship between two populations such as that observed here between the lynx and the snowshoe hare does not *prove* that a cause and effect relationship actually exists. Although the relationship may answer many questions concerning the fluctuations of populations, there still may be another, unobserved factor that is actually responsible.

Many variations in population density can be attributed to climatic conditions. An abundance of rainfall, especially severe winters, or even ordinary seasonal variations are commonly recognized as factors related to fluctuations in populations. In some cases, the relationship appears to be rather clear. But in others, there is widespread disagreement among ecologists. When fluctuations are related to climatic variations or other factors that would be considered to be extrinsic to the population, then they usually do not exhibit regular population peaks.

When fluctuations in populations are related to corresponding variations in the density of the population and the abundance of food (intrinsic factors), they are said to be density-dependent. Such factors seem to regulate vertebrate populations such as fish, birds and mammals, and to some extent, other animals. However, among invertebrates and plants, an irregular fluctuation of population is the rule. The growth of populations and the factors that regulate and control them constitute an area of increasing concern and interest to ecologists. Knowledge of fundamental concepts and relationships existing among animals other than man may provide an insight to problems related to the population explosion of man and measures that can be taken to avert a catastrophe.

QUESTIONS FOR DISCUSSION

1. Describe, in words, what the following formula states in symbols: $D = \frac{N}{S}$.
2. How can we calculate the rate at which a population is increasing or decreasing?
3. What is meant by the terms natality and mortality? Why do they not apply to individuals?
4. What is meant by the biotic environment of an individual?
5. What factors are active in influencing the growth of a population?
6. Describe the characteristic form of a line graph that depicts the growth of a new population living under favorable conditions.
7. Describe the characteristic form of a line graph that represents the changes in a well-established population under natural conditions.
8. How do biologists account for population fluctuations?
9. What are population cycles?
10. Do you feel that mass starvation of animals in nature is common? Explain.
11. Distinguish between immigration and emigration. Explain their importance to population densities.
12. Why is relating the lynx cycle to the hare cycle a reasonable hypothesis? Is there a basis for criticizing the hypothesis that links the cycles of the hare and lynx?
13. Can a valid hypothesis for the population cycle of the showshoe hare be based entirely on diet? Explain.
14. What events may be responsible for the development of cycles in natural populations?

INVESTIGATIONS AND ACTIVITIES

1. Make a report of the work of Malthus concerning populations. In the report, concern yourself with the following questions: (a) did his ideas possess a scientific basis? (b) what were his two basic assumptions? (c) what were his conclusions? (d) what evidence did he use to confirm his theory?

2. Devise and report on the results of an experiment using yeast or bacterial colonies to test Malthus' theory about populations. What justification can you cite for using an experimental study such as this to interpret human population growth?
3. A biologist studied a population of box turtles in an Ohio woodlot for ten years. He determined that the natality averaged 40 per year, the mortality 30 per year, immigration 3 per year, and emigration 8 per year. Using these data, prepare a report of his work, indicating answers and explanations to the following questions: (a) was the population increasing, decreasing, or remaining stable? (b) was the area supplying turtles to other places or vice versa? (c) what was the average annual change due to immigration and emigration? (d) if the initial population was 15 turtles, what must the population have been at the end of the ten years?
4. Obtain the United States census data for your state. Draw a graph of these data beginning with the first census after the state entered the Union. Why is it unnecessary, except in the case of Virginia, to refigure these data for density? How does the form of this graph compare with that for the population of the U.S. as a whole? Attempt to explain any differences, using whatever knowledge you have of conditions in your state.
5. Graph the data of the following table on a piece of ordinary graph paper *and* on a piece of semi-log paper. (a) Has the growth of any of the regions graphed been constant? (b) Which region had the greatest growth rate for the entire period recorded? (c) Which region had the greatest growth rate for the twentieth century? (d) For the decade 1950-60, list the regions in the order of growth rate, with the fastest listed first. (e) As completely as you can, account for the surges in population growth rate of the various regions shown after 1950.

Population (in millions) of the World and Several Regions
Carr-Saunders' Estimates to 1900; United Nations Data Thereafter

Date	Europe and Asiatic U.S.S.R.	Asia Excluding U.S.S.R.	Latin America	World Total
1650	103	327	12	545
1750	144	475	11	728
1800	192	597	19	906
1850	274	741	33	1171
1900	423	915	63	1608
1920	485	997	92	1834
1930	530	1069	110	2008
1940	579	1173	132	2216
1950	594	1272	162	2406
1960	641	1665	208	2972

RECOMMENDED READING

Ardrey, Robert. 1966. *The Territorial Imperative.* Atheneum, New York.

Bates, Marston. 1963. *Animal Worlds.* Random House, New York.

Cahalane, V. H. 1941. *A Trap-removal Census Study of Small Mammals.* J. Wildlife Manag. 5:42-67.

Cole, LaMont C. 1961. *Population Cycles and Random Oscillations.* J. Wildlife Manag. 15:233-252.

Eisenberg, R. N. 1966. *The Regulation of Density in a Natural Population of the Pond Snail, Lymnea Elodes.* Ecology. 47 (6): 899-906.

Hayne, C. W. 1949. *Two Methods for Estimating Populations from Trapping Records.* J. Mammal. 30:399-411.

Herman, Carlton M. 1969. *The Impact of Disease on Wildlife Populations.* BioScience. 19(4): 321-325.

Krebbs, Charles J. 1963. *Lemming Cycle at Baker Lake, Canada, During 1959-62.* Science. 140: 674-675.

Malthus, Thomas R., Julian Huxley, and Frederick Osborn. 1961. *On Populations - Three Essays.* New American Library, Inc. New York.

Niering, W. A., R. H. Whittaker, and C. H. Lowe. 1963. *The Saguaro: a Population in Relation to Environment.* Science. 142: 15-23.

Odum, Eugene P. 1959. *Fundamentals of Ecology,* 2nd ed. W. B. Saunders Co., Philadelphia.

Tabor, R. D. 1956. *Uses of Marking Animals in Ecological Studies: Marking of Mammals; Standard Methods and New Developments.* Ecology. 37:681-685.

Wynne-Edwards, V.C. 1964. *Population Control in Animals.* Sci. Amer. 211(2): 68-74.

5

"There is a procedure in every part of nature that is perfectly regular and geometrical if we can but find it out."
*Woodward**

Community Dynamics

COMMUNITY CONCEPT

Your experiences in the first four chapters should have helped you to formulate some mental images concerning your ecological environment. The use of the term community should evoke a variety of images in your mind. Such images, in all probability, will not be of specific communities, but rather of a more general nature. Mental pictures of this kind are really abstractions of the essential parts of any community. Experiences with a wide variety of communities should enable you to formulate a concept, an understanding of what makes up a community. Even though the abstraction of a community is not of a specific situation in your experience, it should nevertheless contain producers, consumers and decomposers. You will remember that these three elements comprise the first three essentials of a functioning ecosystem. The absence of the abiotic substances distinguishes the community from an ecosystem, adjacent levels of organization in the biological spectrum.

The variety of mental images is reduced if we name a specific type of community; a grassland community, for example. However, we are still dealing with abstract communities, and the image evoked will depend upon our knowledge and experiences with grassland communities. Specific organisms and

*Woodward, Jr. 1699. Phil. Trans. Roy. Soc. London. 21:173.

their respective niches may vary slightly in two similar communities; rarely will two communities be exactly alike. They may, however, share certain characteristics which enable us to generalize, grouping them into a single category. By understanding some of these characteristics, we are in a better position to understand the dynamics of a community, and thereby appreciate and respect the delicate balances which have evolved. The dynamic forces which are continually occurring in a community contribute to that community's stability and integrity.

The community concept is important to understand for several reasons. It enables the average citizen to recognize the importance of conserving natural resources, soil, water, air, and wildlife. These things make his existence more enjoyable and interesting, not to mention that they make it possible in the first place. A trip through diverse communities on a vacation is enhanced if he is able to see behind the scenes and understand the forces at work which make a given community what it is. He comes to realize that in the great diversity of organisms there is an intricate orderliness. Plants and animals do not just happen to be where they are. Their adaptations to a particular community situation are the result of many hundreds of generations of their ancestors interacting with not only other members of their own species, but also with other species as well as the abiotic features of their environment.

The community concept is also important to the average citizen as well as to the ecologist in another very significant way. The various news media are devoting more and more attention to the importance of problems imposed by man upon the environment. The public must not only be aware of such problems, but must understand why they exist, and the variety of means that are available to cope with them. Furthermore, he must understand that the simplest and least expensive way to overcome such problems is not necessarily the simplest and least expensive in the long run. The use of pesticides in controlling insects has been a boon to health and agriculture since their introduction in the early 1940's. However, we are only beginning to understand some of the negative ecological consequences of widespread and often indiscriminate use of such tools. Mosquitoes, for example, can often be more effectively controlled by environmental modification than by direct attack poisons.

Other examples, such as the predator control program in the Kaibab National Forest cited in the last chapter, emphasize the need for understanding communities and the dynamics which control them. Ultimately, we, as citizens of the ecological community, must depend upon healthy natural communities. Consequently, it is important for us to realize and understand that even insignificant acts that have seemingly become part of our culture have their ecological ramifications.

ECOTONES

Occasionally, where abrupt changes occur in the geography of an area, an equally abrupt transition from one community to another may occur. More often, however, the lines separating different communities are not clear, and a gradual transition zone from one community to the next is apparent. Such a zone is called an *ecotone* (Fig. 5-1).

The greater the differences between two adjacent communities, the more apparent is the ecotone between them. The width of the ecotone community depends upon many factors, but is usually restricted to a relatively narrow belt between two larger, adjacent, well-defined communities. The ecotone area itself represents a zone of tension in which varying climatic conditions, the intervention of man or other factors begin to limit the spread of a given community. Within the ecotone are commonly found many of the organisms of the two adjacent overlapping communities. Many other species are well adapted to the conditions of a transition zone, and will even be restricted to such areas. Because of the blending of characteristics of the two communities, the number of species and occasionally the population density of some of these species will be greater within the ecotone than in the communities on either side.

Why are more species usually found in a transition zone? For the answer to this, we must return to a previous discussion: that of limiting factors. Many species will have a relatively wide degree of tolerance to the new environmental conditions which are encountered on the edges of a community. As a result, they will tend to invade adjacent communities. Not all

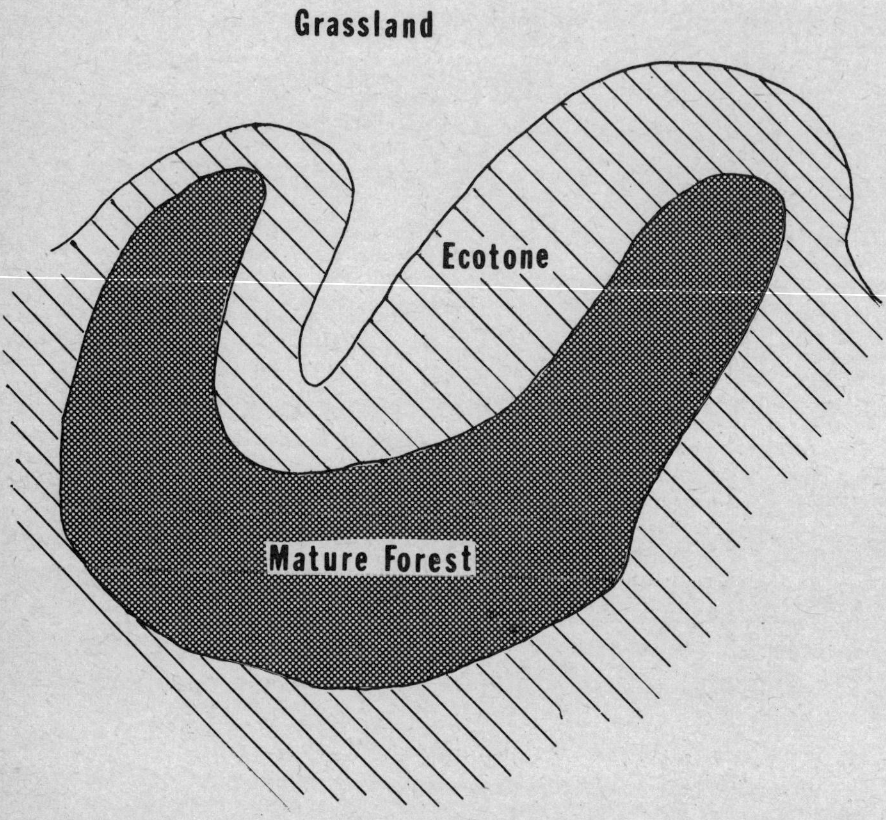

Fig. 5-1. A mature forest surrounded by a grassland community. The ecotone area separating the two communities is of variable width and typified by scattered trees and the more shade-tolerant grasses.

An ecotone between an open field and a forest is a blending of the two communities. Small trees, shrubs and grasses provide a habitat in which some of the organisms from both communities will visit as well as providing a home for animals that are especially well-adapted for the conditions of the ecotone and not for either of the adjacent communities.

species will be tolerant to the new conditions, and their success as an invading species will be limited at best.

Perhaps one of the most obvious ecotone communities is that encountered between forest and prairie communities. Trees with a relatively high tolerance for direct sunlight, warmer temperatures, and decreased moisture will invade the adjacent prairie. At the same time, some of the grasses which can survive with limited direct sunlight and cooler temperatures will spread into the forest edges. Examination of an ecotone area of this type reveals a few scattered trees, probably smaller than trees of the same species in the forest, scattered shrubs and a ground cover of the more tolerant prairie grasses. Compared to the adjacent communities, conditions of the ecotone are relatively unstable. This situation provides a favorable environment for new species, unique to the ecotone, to establish themselves. The new community within the transition zone then becomes a special situation. It possesses characteristics of the adjacent communities, but it also develops characteristics of its own.

The tendency for communities to blend together in this manner has sometimes been referred to as the *edge effect*. This principle has been put to practical use in recent years by lumbering concerns. Instead of removing large blocks of trees and then replanting in the cleared areas, more and more lumbering is done in strips. The cleared areas are then planted to hardy, dense cover species such as blackberry. The shade provided for young tree seedlings by the dense bramble is extremely favorable for the invasion of the adjacent forest species. The return to a stand of trees for lumber is much more rapid than would be possible by using previous methods. We will return to this principle from a different point of view later in the chapter.

The ecotone area will be relatively narrow where the physical factors change abruptly. This is the case where an aquatic community such as a marsh, a pond, or stream imposes itself upon a terrestrial community. A stream or pond will often be flanked by plant species such as cattails and willows which then give way to the surrounding community a short distance from the water's edge. Associated with these transition plants will be animal species which are characteristic of such an area. Streams

Lichens represent the pioneer species on rock surfaces. Their presence helps in the decay of the rock and the build-up of soil materials in the lower recesses of the rocks. These recesses will retain some moisture, and other plants will invade.

and rivers cutting through the vast expanses of the prairie states are easily detected by the long fingers of dense wooded areas lining the banks on either side. Left undisturbed, these river-bank species will include cottonwood, oak, American elm, walnut and hickory trees, as well as numerous shrubs and vines. Certain faunal species, such as raccoon, opossum, a variety of game and song birds, and deer will be encouraged by the narrow band of ecotone between the water community and that of the surrounding cropland or prairie. The line between this area and cultivated land which may border it will be especially abrupt since the boundaries here are artificially maintained.

Man is usually identified as a forest-edge species. Evidence for this may be seen in the history of the American settler as he moved across the continent. When he settled in the forests, he cleared them, leaving a few trees scattered about, interspersed with grassland and cropland. When he settled on the prairies, he planted trees, using them for windbreaks for his dwellings and shelter belts for his stock and game animals. Since he is in the constant process of creating edge areas, he is constantly encountering plants and animals which are typical edge species, and many of which he considers as pests. Many of these species, including weeds, insects, and other "campfollowers," have been able to greatly expand their ranges as a result of man's alteration of the environment.

COMMUNITY SUCCESSION

Perhaps one of the most dramatic transactions that is characteristic of a community in action is that of *succession.* Ecologically, succession is the orderly replacement of one community by another over a period of time. Sometimes the process occurs at such a slow pace that visible alterations in a community are not detectable during one person's lifetime. But this is not always the case. A succession of several communities can be observed in a classroom aquarium over a period of a few months. Starting with a volume of boiled pond water, to which a handful of dried grass and weeds from a roadside ditch has been added, you can chart the progression of organisms and their relative abundance.

Where do these organisms come from? What are the first organisms to appear? Does one kind of organism seem to be replaced by other kinds after a period of time? What factors contribute to the apparent succession of communities in the aquarium? Can definite points be established which separate one community from the next? These are questions that should guide your thinking as you observe a microcommunity of this type in action.

There are two basic types of succession that occur. *Primary succession* describes the process which occurs when life becomes established in a particular area for the first time. There are several areas where this might occur. Bare rocks, newly bared soils due to landslides or other natural processes, volcanic ash, newly formed islands in the oceans, or sand dunes provide substrates on which no life has ever before become established. The first species to colonize a new surface are called the *pioneer species.* The pioneer community will slowly alter the conditions on that substrate and thereby pave the way for new species to become established in the community. The development of *secondary communities,* in turn, alters the conditions with two consequences. First, conditions for the primary community to survive become so altered that its members die out or leave. Second, members of a secondary community continue to change the environment enabling other species to colonize. Thus, the process of one community replacing another occurs again.

A given community which becomes established for a period of time during the process of succession is called a *seral stage.* Collectively, all of the communities involved in the progression will be termed a *sere.*

Secondary succession is the other basic type of succession which can occur. In this process, succession does not begin in a basic environment such as described above. Various communities have already existed in the area, and have even progressed through many seral stages. However, the process of succession in the area has been interrupted by some relatively catastrophic event such as a fire, violent weather conditions, or the intervention of man through lumbering, farming, or pollution. Occasionally, the succession which had been in progress is merely set back to a previous seral stage, and the progression

Where succession is disrupted and is forced to begin again in an area, the process is called secondary succession. Adjacent communities may remain undisturbed.

of communities begins again. This is possible because the area may be relatively small, and seeds from nearby areas can quickly become established.

On the other hand, the sere may take on an entirely new pattern. Although the end result may be the same, the sequence of stages in its development may well be different. This is true primarily because of the availablity of plants and animals which can become established in the disturbed area which, in all probability, were not available during the earlier stages of primary succession. Also, many conditions, established by the communities before the sere was interrupted have been sufficiently altered to allow a different series of communities to colonize the area. The minerals present in the soil, for example, and the water-holding capacity as well as many other features will provide an entirely different set of abiotic factors which will dictate the course of secondary succession.

How long does the process of succession take? This question implies that there is an endpoint at which the process stops. Although biologists may not totally agree, such an endpoint generally exists. When a particular seral stage is able to perpetuate itself indefinitely with no significant changes in the flora or fauna of the community, the process stops. Such a community is known as a *climax community*. It is in equilibrium with the climatic factors of the region. The time required for a climax community to be attained is dependent upon several factors.

Temperature ranges, relative humidity, wind and precipitation are factors which are important in determining the climax community for a given area. Two entirely different areas with respect to these factors may go through the same seral stages up to a point. One or more of these factors may become limiting for the next stage to develop in one of the areas, and the sere, having reached its climax community for that area, ceases further development. The other area, not being limited by those factors, will continue to progress toward a different climax community. When climate or other factors begin to be a limiting factor for this second area, further succession will cease here, too.

Community dynamics do not stop at this point, however. On the contrary. There will still be the continual transfer of

The beech and maple forest found in the eastern portion of the United States is an example of a climax community. It is in equilibrium with the climatic factors of the region.

energy through the various trophic levels. There will still be population fluctuations in response to innumerable pressures. Only the replacement of one community by another actually stops. The natural community dynamics will enable this final seral stage to survive as long as the climate and physiology remain constant, and as long as it is undisturbed by man.

There are two concepts regarding successional changes which must be kept in mind when discussing the events which take place. First, although we generally refer to various seral stages through the use of the flora involved, we must be aware of a corresponding succession of animals, or fauna. Floral changes are usually the most obvious and therefore useful as points of reference. Second, changes in the types of vegetation will usually have the greatest impact upon the environmental situation. Can you think of reasons why this might be so?

Climatic Succession. It is certainly no secret that the climate of the world is constantly changing, even if slowly. The presence of coal deposits in Antarctica give ample evidence that tropical conditions once existed on that now frozen continent. Although many theories have been advanced to account for such a phenomenon, the point is clear that the climate is now far different from what it once was. In the northern hemisphere, the periodic advance and retreat of the polar ice cap has produced a series of climatic alterations that will be reflected in the types of communities that can develop in particular areas. Fossil plants can be identified not only by their leaves and flowers whose impressions are readily observed in certain rock strata, but by the particular structure of its pollen grains. Through fossil techniques, we are able to know the types of trees which existed at that location as much as 10,000 years ago or more. Knowledge of the flora of the region will thus give us an indication of the climate.

Indiana bogs, for example, were gouged out by the last glacier and then left as the glacier melted during the ensuing warming period. Sediments at the bottom of the bogs contain pollen grains describing the succession of trees that surrounded the bogs as the climate continued its warming trend. The lowest layers contain pollen of the balsam fir, indicating a cool, moist climate. Later deposits contain a predominance of pollen from spruce, indicating a drier, but still cool climate. As the temperatures warmed and the land dried still further, pines apparently moved into the area. These were finally replaced by the oaks, characteristic of the present warm, dry climate of that area. Where pollen is insufficient to establish the course of climatic succession, the presence of fossilized portions of plant and animal skeletons is also used as an important tool. Figure 5-2 summarizes the sequence of events involved in climatic succession.

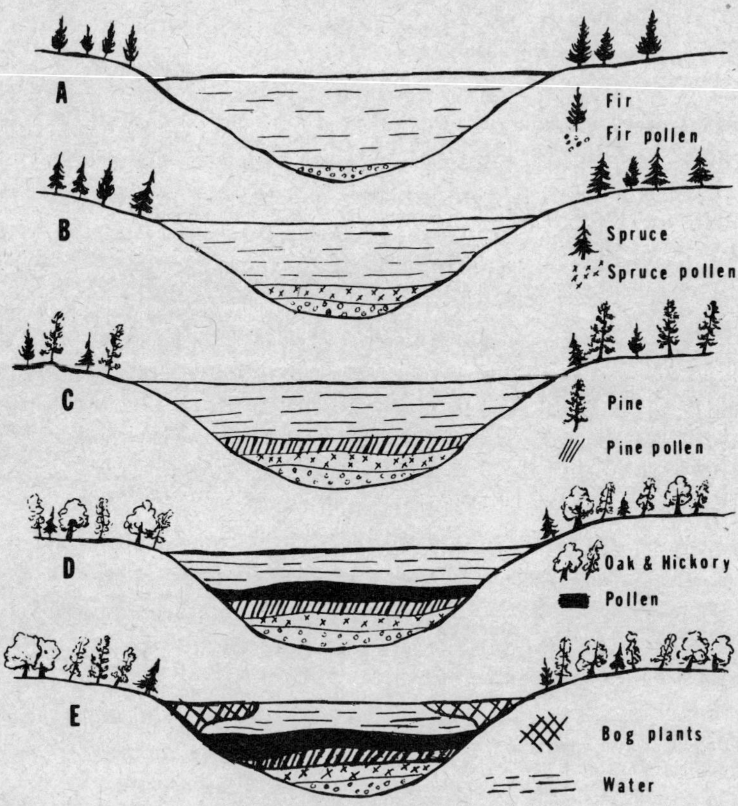

Fig. 5-2. Steps involved in climatic succession as revealed by pollen deposits in bogs. Balsam fir (A) was the first tree species to appear after retreat of glacier. (B) As the glacier retreated still further, the fir moved north as white spruce came in. (C) Spruce moved northward as climate continued to warm and was replaced by pine. (D) Milder climate favored growth of oak and hickory, replacing pine. (E) Some earlier species remain on the bog mat which will eventually cover the lake. (From Hylander, 1966.)

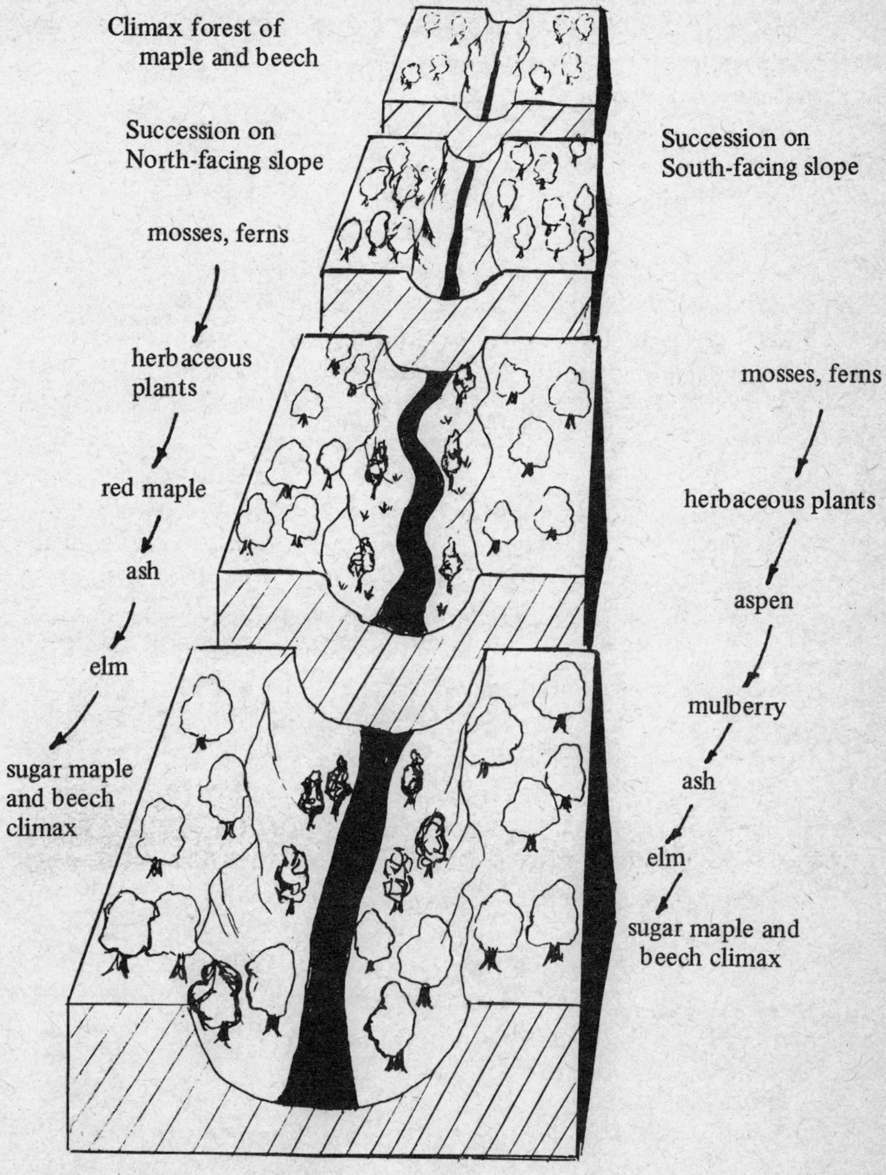

Fig. 5-3. Physiographic succession as caused by a stream eroding away its banks in an eastern beech-maple climax forest. (From Hylander, 1965.)

Physiographic Succession. Changes in the surface features of the earth, not necessarily associated with climatic changes, may also be a fundamental cause of succession. It may occur in an area where a climax community has already been established. Its cause may simply be the eternal chipping away of land by a stream as it courses its way to the ocean (Fig. 5-3).

In a climax forest, as the stream erodes away its banks, a wider stream bed with steeper banks develops. The moist banks are well-suited for mosses and ferns which are the first to colonize the newly exposed ground. Trees lining the banks are slowly undercut by the stream, and they eventually fall, widening the stream even more. Eventually, as the stream matures and cuts a more and more meandering path through the area, a broad valley develops. Few trees, of course, are found in the valley. The warmer, drier conditions in the valley act as limiting factors preventing the spread of the nearby climax species into the bottom land. Light-tolerant trees, shrubs and grasses invade the broad valley with the inadvertent help of wind and animals, and eventually create environmental conditions which encourage the spread of the climax species down the slopes and onto the valley floor. As they in turn alter the environment, the original occupants must give way and be replaced until finally, the climax community dominates the new area.

Biotic Succession. As has been pointed out already, given stable climatic conditions, unchanged physiography, and protection from the effects of man, the climax community for an area will always be the same. It does not matter whether the starting point happens to be a bare rock surface, a newly formed island, an area of sand dunes, or a pond. To illustrate the sequence of events in biotic succession, let us consider the development of a climax community beginning with opposing conditions: the bare rock and the pond.

The rock surface presents a difficult set of conditions for colonization by plants. There is no soil; temperatures usually have a wide range from the coldest to hottest extremes; there is little, if any, protection from the wind; and to top it off, moisture is scarce. Nevertheless, this area can still be colonized by a primitive plant form, the *lichen.* There are many types of lichens, but the first to colonize the bare surface are crust lichens. These are low-lying forms, occurring in a wide variety of colors. They attach to the substrate by small root-like structures called rhizoids. Secretions from the rhizoids help to dissolve the rock surface, breaking off small pieces. Freezing and thawing of ice in the small openings and cracks in the rock, coupled with the slow dissolving action of water also helps to break the rock into smaller particles. This represents the first step in soil-building.

The cracks and crevices opened up by weathering and the action of the lichens provide a place where dust particles and various other forms of organic and inorganic debris can collect. Thus, a shallow soil surface is formed in which moss spores and grasses tolerant to relatively dry conditions can germinate. As these plants take hold in the new community, they cover the original lichen community, adding to the thickness of the forming soil. New environmental conditions begin to emerge. The temperature range experienced in the new community is reduced by a biotic cover on the rock surface. Wind velocity at the surface is reduced as grasses form a windbreak. The soil retains some of the moisture that it obtains from rains and run-off. Such environmental alterations provide conditions which favor other plants. Soon, dry-tolerant trees such as spruce and pine are able to gain a foothold. The strong roots, penetrating the substrate in search of moisture exert enormous pressures upon the surrounding rocks, breaking them into smaller pieces and aiding in the soil-building process. These trees will eventually be replaced by fir, oak and hickory, and finally, under ideal conditions, beech and maple as climax species (Fig. 5-4).

Prevailing climatic conditions may prevent the attainment of the beech-maple community. Consequently, an oak-hickory climax may represent the final dominants. As in the case of many parts of the Rocky Mountains, spruce-Douglasfir or a ponderosa pine forest may be the final seral stage. Other climax communities include the pinon pine-juniper climax found commonly in the high deserts of Utah, western Colorado and Idaho, as well as an ecotone between the grasslands and the mountains in Montana, Wyoming, Colorado and New Mexico. The Great Plains illustrates the culmination of a sere in grass as the climax species. The tall grass prairie in the eastern portion of the Plains, receives more annual precipitation and is

The breakdown of large rocks may be accomplished by alternate expansion and contraction due to heat and cold. Moisture, entering small cracks, may freeze and expand thereby increasing the size of the crack. Ultimately, the rock is reduced to smaller fragments.

Type of Animal	Name
Mammals	Gray fox Raccoon Eastern chipmunk White-footed mouse Eastern mole
Birds	Whippoorwill White-breasted nuthatch Ruffed grouse Acadian flycatcher
Reptiles	Timber rattlesnake Copperhead Black rat snake Skinks Box turtle
Amphibians	Marbled salamander Red-backed salamander Wood frog Common tree frog

Table 5-1. Some of the Common Vertebrate Representatives of an Oak-Hickory Climax Forest.

made up of bluestem grasses, wheatgrass, dropseed, and others. In the west, these grasses gradually give way in the face of drier conditions to the short grass prairie. Dominant grasses here are gramas and buffalo grass.

Paralleling the development of these vegetational communities, new niches will develop and old niches will become obsolete. As a result, there will be corresponding succession of animals with characteristic animals for each stage. Table 5-1 summarizes the important fauna of a typical oak-hickory climax community.

The succession of communities from a pond to a climax community presents an entirely different sere. The time required for the process is usually considerably less than that required for succession beginning on bare rock. Except for large lakes, most ponds are relatively transient features of the landscape. They may be products of natural forces such as glaciers, or they may be man-made. Given enough time, without interference, their fates are similar: replacement by a climax community.

The surrounding area which drains surface water into the lake is called the *watershed.* But water is not all that enters the pond. Soil and other organic matter is also carried into the pond and is deposited on the bottom. Pioneer vegetation in the pond is made up of bottom-rooted pond weeds such as Elodea and tapegrass. The build-up of sediments contributed by the watershed and dead water plants shallows the pond and enables floating plants rooted in the bottom to replace the bottom-dwellers. Dense mats of such vegetation around the edges of the pond aid in trapping sediments and further shallow the pond. A marshy perimeter develops around the pond, ideal for the emergence of cattails, rushes and sedges. They, in turn, contribute to the build-up of soil, and the marsh community gives way to a drier community of woody shrubs, willows and the like. As the land dries out even more, these forms give way to the forests, eventually reaching the climax community dominated by beeches and maples (Fig. 5-5). As in bare rock succession, climatic conditions may limit the attainment of this particular climax community, and succession will stop at a subclimax stage such as has already been mentioned.

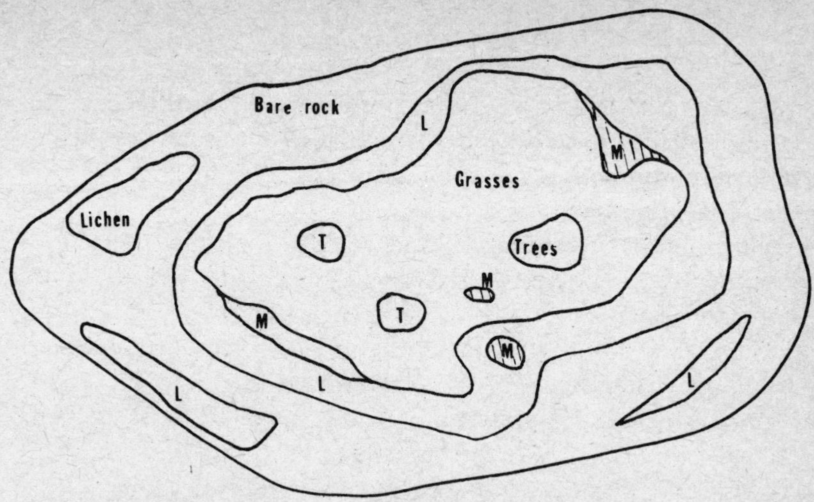

Fig. 5-4. Bare rock succession. Note that advanced stages (trees, T) will be located where the newly formed soil is the thickest. Pioneer species (lichens, L, and mosses, M) surround the successional zone.

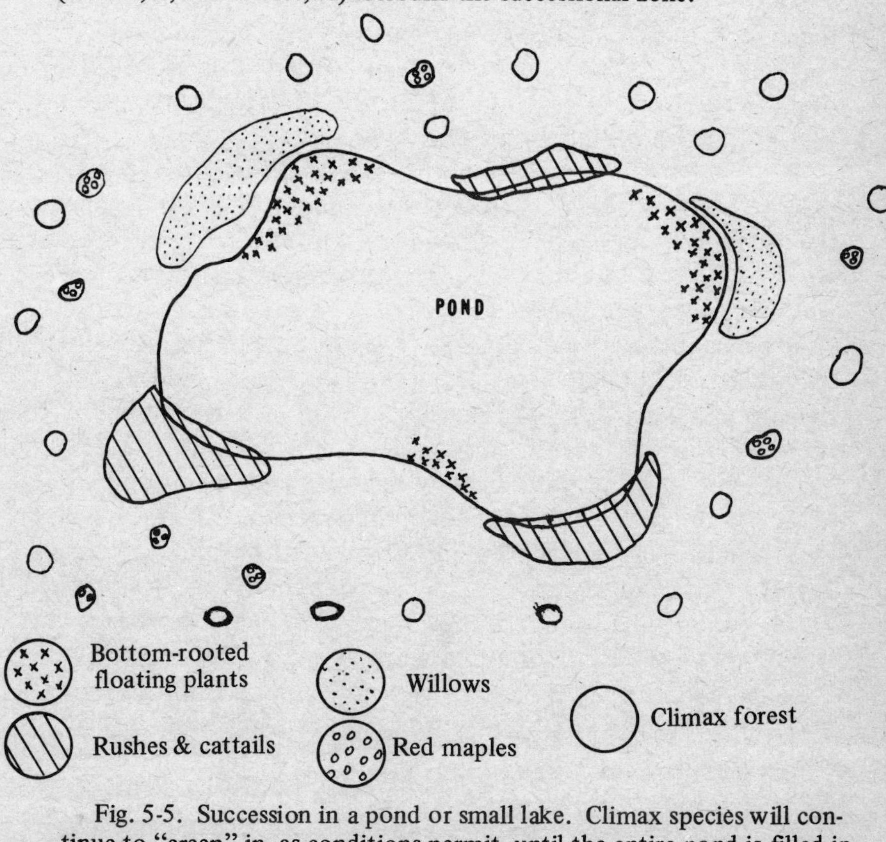

Fig. 5-5. Succession in a pond or small lake. Climax species will continue to "creep" in, as conditions permit, until the entire pond is filled in and becomes part of the climax community.

It should be emphasized that time alone is not the controlling factor in succession. The climate, soil and topography of the particular environment ultimately govern the rate and climax stages of a sere. In some areas, succession occurs relatively quickly, reaching a climax community in two hundred or so years. In other environments, particularly where moisture is scarce, little noticeable succession may occur in the same time span.

Many areas have felt the unwise hand of man upon perfectly good climax communities. Perhaps one of the most lucid examples of this is found in the short-grass prairies of the Great Plains. Buffalo and pronghorn had survived on these prairies and the taller prairies to the east for hundreds of years. Climatic conditions in the eastern portions were not so limiting and trees would continually invade the lush grasslands. However, the Indians, in order to preserve the vast grazing land for the game which sustained them, learned that an occasional burning would prevent the next stage of succession from occurring. In the western portions, preventive maintenance of the prairie was not necessary.

The advent of the white man to the Great Plains was a catastrophe in many respects. Most significant was the removal of the buffalo. Ranchers replaced these natural grazers with cattle. Whereas the buffalo had a free range and remained in one place for only short periods, the ranchers continually maintained large herds on the same range. It was not long before the climax community had been set back far enough that it could no longer maintain itself. The range was badly overgrazed and there were no plants to hold the soil. The grassland fauna, no longer able to survive, vanished with the grasses. Bare soil, dotted with sagebrush became a common sight. The crowning blow came in the period 1933-1939. Spared the final catastrophe until then by more than average precipitation levels, the ranchers continued to press their luck. But the Dust Bowl years were to have the last word, and when the rains and snows failed to come, the well-known wind of the plains finished the lesson by blowing away the thin topsoil, leaving total devastation in its wake. It will be many years before the climax community is again able to become established in these area. Man does not make it any easier. Claiming that improved ranching

Beaver may also effect changes in the environment through his dam-building. The shallow ponds become shallower over a period of time, and the ponds become marshes. Finally, a stream courses through what had been the beaver's home, and the filled-in pond ultimately becomes a climax forest.

and farming methods along with irrigation make the Dust Bowl days obsolete, he has gone back to the same practices of laying the soil bare and overgrazing limited land.

Herbert H. Ross, principal scientist of the Illinois Natural History Survey has summarized the process of succession:

> The same mechanisms of succession occur in all of these examples. Throughout any one series of community successions, the seeds of all plants from all the communities reach or fall on the area occupied by all the communities. Seeds of climax species falling on areas in the early stages of succession either will not germinate, or the seedlings die. . . . Seeds of subclimax species falling in the climax area have the same fate. However, when the organisms growing in one of the subclimax communities have either added sufficient humus to the soil, or produced sufficient shade, or brought about some other critical effect, then the progeny of the organisms in the next successional stage can reproduce and grow better than those of the community occupying the area. This change then eliminates that particular subclimax community in that area.

Farmers must wage a constant battle against the forces of succession. Grasses, weeds and woody seedlings are constantly invading cleared cropland and over-grazed pastures. Abandoned farmsteads aptly illustrate the re-establishment of a successional pattern. This is especially evident in the eastern half of the United States. Abandoned pastures, protected from periodic fires and cultivation are supporting seral stages which will ultimately culminate in the climax community for that region. The process can be observed wherever chemicals, fires, or over-grazing have modified or substantially altered a natural successional stage.

There is not much that man can do to hasten the process of succession. Earlier in the chapter, it was pointed out that strip lumbering had shown better results than reseeding large tracts with the climax species directly. Replanting burned-over forests in climax species has not been as satisfactory as had been hoped. We are even becoming aware of some of the natural processes involved in the maintenance of a climax community – processes that are not at once obvious. For example, natural forest fires which occurred periodically throughout the forests of North America even before the advent of the white man's invasion, had an important function. The fires prevented the build-up of large amounts of organic debris on the forest floor, and kept the forest relatively clear. Fires, when they occurred, had less to burn, and therefore, were not as hot. They did not reach the upper levels of the trees. Forests were able to survive these low-grade fires with only a minimum of interruption. Today, although fires do occur, they are generally less extensive. Litter continues to accumulate. When a big fire develops, it burns hotter and often becomes a crown fire, resulting in more serious destruction than would occur if the forest was not so "protected."

SEASONAL FLUCTUATIONS

Community fluctuation refers to the alterations in activities of the organisms in the community which result in regular, periodic changes in the character of the community. Throughout most of the world, seasonal changes will produce fluctuations in the community character which are readily observable. It should be somewhat apparent that community rhythms are actually the result of rhythms in the individual populations which make up the community.

In terrestrial environments, there are two principal factors which control seasonal fluctuations. First, the temperature is perhaps the most obvious. You are no doubt quite familiar with the changes in temperature associated with the changing seasons. The second factor, although equally familiar, is perhaps not as obvious to you as a factor which is responsible for changes in community activities. Nevertheless, the *photoperiod* (relative length of day and night) has been the subject of much research in recent years. In many environments, especially those of the tropics, seasonal alterations in rainfall or relative humidity will be instrumental in altering the seasonal complexion of the community.

In marine (ocean) environments, temperature fluctuations and periodic alterations in salinity (salt concentration) are the primary causative agents in seasonal fluctuations. What factors can you think of which might result in regular changes in the salinity of parts of the ocean?

It is probably not often that a single causative agent is responsible for the alteration of a community's complexion as seasonal changes occur. In fact, several such factors may work in combination to produce the observed effects. The factors mentioned thus far have been of an *extrinsic* nature. That is, the stimulus which produces a change in the activity of an organism occurs outside of the organism.

There are undoubtedly also a great many *intrinsic* factors yet to be discovered. Some organisms are capable of exhibiting fluctuations in activity which are regular and apparently independent of any external stimuli. A great many organisms fall somewhere in between. They possess some rhythmic activities that are coordinated with changes in the physical factors of their environment, but when these stimuli are removed, the rhythmic activity continues, at least for a time. There are definite advantages to organisms which possess intrinsic rhythms coordinated with changes in the environment. Such organisms are able to undergo various physiological adjustments in preparation for changes in climate, an advantage that contributes to the survival of the organism.

Seasonal temperature fluctuations occur throughout most of the world. Only in deep caves and certain lakes (as well as the ocean benthos) are seasonal temperature changes relatively unimportant. The rhythmic rise and fall of temperatures through the seasons is a result of the movement of the earth around the sun (Fig. 5-6). When the northern hemisphere is in the midst of its winter, the sun's rays shine almost directly upon the southern hemisphere. Although the distance between the earth and the sun is least at this time, the important factor is the directness of the solar radiation reaching the earth and not the distance it travels. The amount of solar radiation falling on one square foot in the southern hemisphere is spread over a larger area in the northern hemisphere due to the curvature of the earth. Since less energy per unit area falls upon the northern hemisphere during winter, its effects will be limited accordingly. There will be less heating, less photosynthetic activity, and as a result, there will be alterations in the character of the northern communities in response to such environmental changes.

You can demonstrate the effects of the earth's curvature using a piece of cardboard and a flashlight. Holding the cardboard vertically, shine the light directly upon it at a right angle to the surface. Notice the area of the spot of light. Then tilt

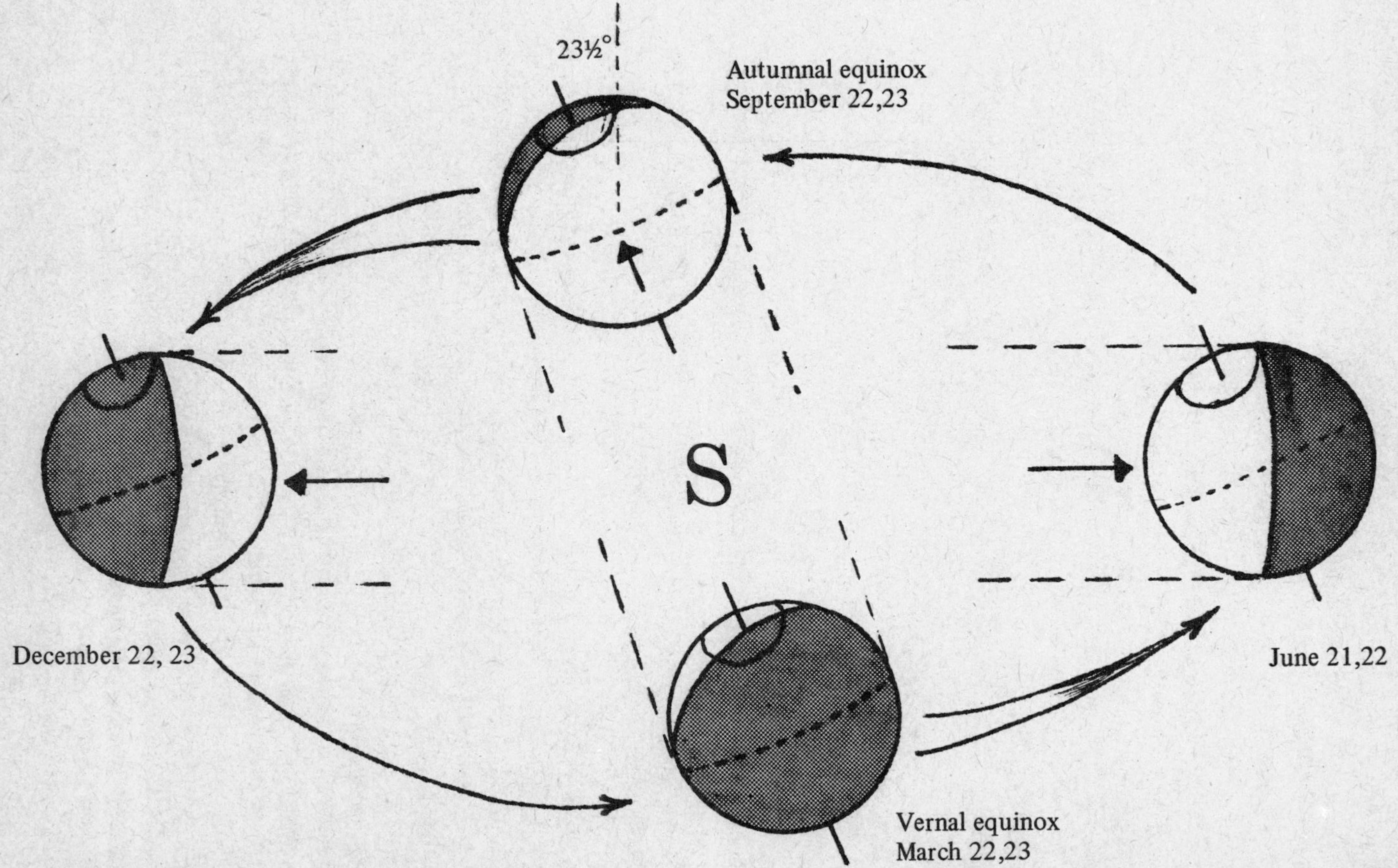

Fig. 5-6. The inclination of the earth's axis remains fixed at 23½°. As the earth revolves about the sun, different portions of the earth receive the most vertical solar radiation at different times of the year, causing the seasons.

the cardboard back, keeping the light in the same place. Notice how the area of the spot of light increases. Now the same amount of energy is falling over a larger area, and the intensity is reduced accordingly.

Summer in the northern hemisphere occurs as the sun's rays become more direct. The change in the angle with which the solar radiation strikes the earth is brought about by the combined effects of two factors. First, the axis of the earth is inclined at an angle to the plane of its orbit. This angle is unchanging. Second, the earth continues to revolve about the sun. Together, these factors bring different areas of the earth under the most direct rays of the sun in a regular, periodic manner, producing the familiar seasons. The evolutionary processes at work in the biotic world have produced adaptations in organisms which are stimulated and coordinated with these seasonal fluctuations.

The photoperiod is also a changing environmental feature in response to the earth's movement about the sun. Furthermore, it is an extremely accurate "time clock," being constant year after year. It is no secret that the northern hemisphere experiences longer nights and shorter days during the winter. As spring approaches, the days grow longer, radiation becomes more direct, and temperatures rise.

The changing photoperiod is known to produce a wide variety of changes in the biotic community. Migratory flights of birds are perhaps one of the most well-known examples of animal displays in response to changes in the relative lengths of days and nights. The photoperiod is also implicated in other seasonal activities such as breeding, hibernation, estivation, and the development of the condition of the pelt or plumage of some animals. Particularly noteworthy as examples of the latter situation are the seasonal changes of pelt coloration of the weasel, snowshoe hare and the ptarmigan, all of which turn white in the winter months. The white coat is a protective coloration against the snow-covered environment. When these animals are kept in an environment of long days and short nights, such as would be found during the summer months, the white coat fails to develop "on schedule."

Less well known are the responses of plants to changes in the photoperiod. Most plants will flower in a specific part of the growing season. The flowering process is known to be a direct response to the specific length of darkness which occurs at its flowering time. Laboratory experiments have shown that flowering in such plants can be initiated and maintained by artificially controlling the length of the dark period. Some plants, such as the soybean and chrysanthemum are extremely sensitive to interruption of the dark period by a short exposure to a light source. In some cases, the flash of a match or a flashlight during the critical period of darkness is sufficient to inhibit the flowering process entirely! Some plants need a certain number of successive light-dark periods of a critical length for flowering to be initiated. Others need only a single night of suitable length to bring about flowering. Interruption of the dark period for these latter plants might be overcome the following night, providing they were not disturbed again.

LUNAR RHYTHMS

Many marine organisms exhibit rhythms related to the ocean tides which are produced by the combined effects of the moon and sun. If you live near the seashore, you are undoubtedly familiar with the daily rise and fall of the ocean. Noticeable higher tides (*spring tides*) occur twice each month when the gravitational effects of the sun and moon supplement each other (Fig. 5-7). Similarly, when their effects tend to cancel each other, the daily high tide is at its lowest point. This is called the *neap tide*.

Marine animals living in the *intertidal zone* (the area between the highest and lowest tides) will experience more noticeable effects from tidal changes than animals living in the open seas. Many become inactive during low tides when they are exposed to the atmosphere and direct solar radiation. Food-getting, and other activities must be limited to periods of high tides when they are covered with water. Food production by algae found in the intertidal zone is also limited to periods of high tides. Terrestrial animals exhibit little or no rhythmic activity which can be positively correlated with the lunar cycle.

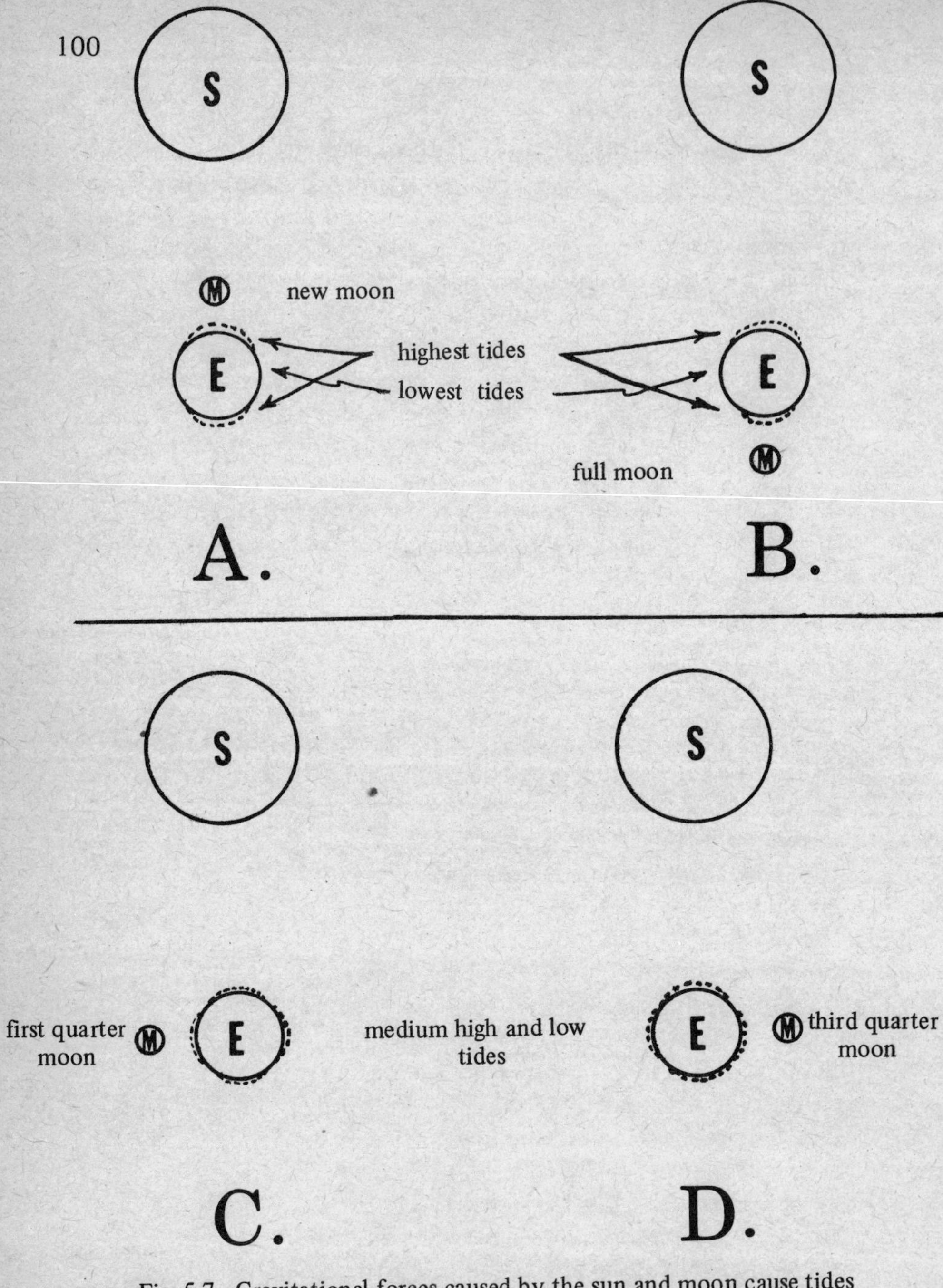

Fig. 5-7. Gravitational forces caused by the sun and moon cause tides to occur on the earth. During lunar phases of new moon and full moon (A and B), sun and moon supplement each other causing highest tides. During first and third quarters, their gravitational forces work at right angles to each other (C and D) causing only medium high and low tides.

The intertidal zone is a relatively harsh environment being alternately exposed to the strong forces of wave action and the drying effects of the sun during low tide.

In terms of the longer lunar cycles, that is, the monthly spring and neap tides, reproductive activity of the grunion is well known. During the highest tides, millions of these tiny fish are washed ashore along the southern California coast. The female buries herself in the sand and deposits the eggs which are then fertilized by the male. With the next wave, the fish are washed back out to sea, leaving the buried eggs to develop. During the next spring tide which occurs two weeks later, the eggs hatch and the young grunion are carried to the sea.

CIRCADIAN RHYTHMS

Every twenty-four hours the earth rotates once on its axis. Consequently, there are alternating periods of light and dark, the basis for *circadian,* or daily rhythms. The earliest clear and extensive demonstrations of the presence of a clock system in organisms came from the studies of young bean seedlings. Wilhelm Pfeffer, a German plant physiologist, became interested in the so-called sleep movements exhibited by some plants. During the day, the first pair of leaves of the young bean seedlings would be elevated. At night, these leaves would drop a noticeable distance. Maintaining the plants under constant light and temperature conditions indicated that neither factor was responsible; the sleep movements persisted. The rhythm could be reset, however, by subjecting the plants to darkness during daylight hours and to illumination during the night.

Many flowers are also subject to daily rhythms. Some are well known for their nighttime flowering. Others open in the early morning only to close a short time later and remain closed until the following morning.

Animals, including man, are closely geared to the daily pattern of the 24-hour day. The American cockroach is a nocturnal (night active) insect with its greatest activity starting an hour or so after dark and continuing for five or six hours. A specific rhythm can be established in these insects by subjecting them to alternating 12-hour periods of light and dark. When they are then transferred to a condition of continuous darkness, they become active slightly earlier each day. When they are transferred to continuous light, there is a delay in activity starting time. There is apparently some intrinsic mechanism that is closely coordinated with the daily rhythm of the physical environment.

The white-footed deer mouse is another example of an organism with an intrinsic circadian rhythm. Also being nocturnal, its activities, including feeding, running, and waste removal are all geared to the hours of darkness. Other mice are not limited to specific hours of activity. The pine vole, for example, alternates periods of activity and rest throughout the 24-hour day. Figure 5-8 illustrates graphically the comparison of the activity periods for these two animals.

The jet age has demonstrated that man is also subject to circadian rhythms. When he travels across several time zones in a short time, it often takes several days to become readjusted to new patterns of eating, sleeping, and being active. Common physiological effects of circadian rhythms can be measured easily. Body temperature fluctuates regularly as does the volume of urine produced.

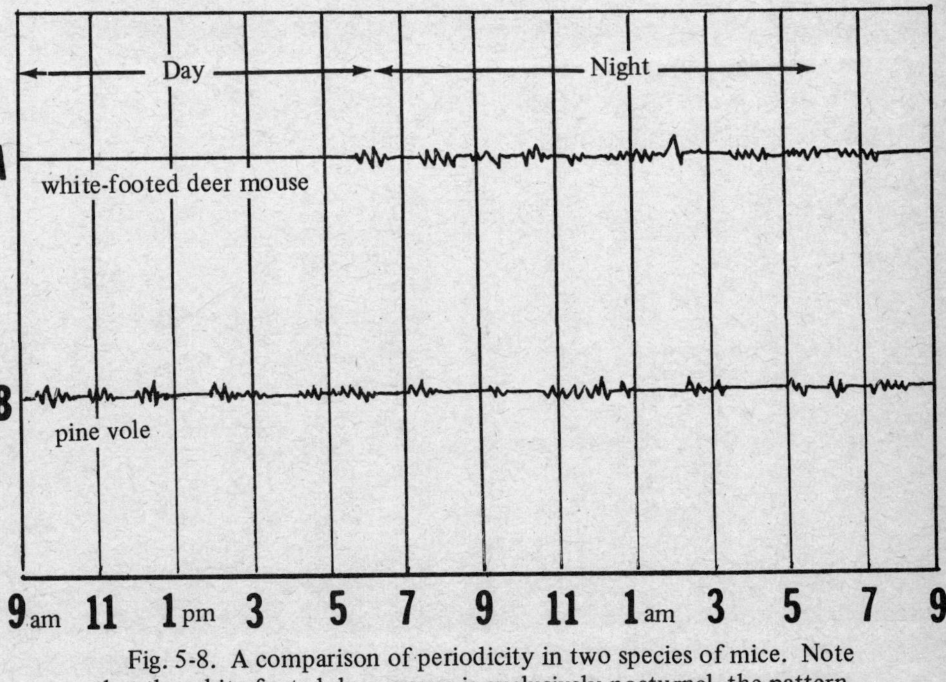

Fig. 5-8. A comparison of periodicity in two species of mice. Note that the white-footed deer mouse is exclusively nocturnal, the pattern persisting in continuous darkness. The pine vole alternates activity and inactivity throughout the day in spite of continuous darkness. (From Odum, 1959.)

ARHYTHMIC FLUCTUATIONS

The progression of seasons, the regular alternation of days and nights, and the lunar cycle are all periodic and predictable. Because of their regular nature, much can be learned about the rhythms of plants and animals that are closely associated or at least correlated with these environmental rhythms. Many events, however, are not regular and predictable. They nevertheless have a great deal of influence upon the character or complexion of a community at any given time.

Climatic factors are extremely important to a community. They not only influence the progress of succession, but they may temporarily alter the state of any given community. Two successive years are seldom identical with respect to the weather. We often speak of "wet" years and "dry" years or "hot" summers and the like. Especially abnormal rainfall, drought, temperatures, or humidity may be felt in a community by interrupting regular rhythmic activity.

Fires are also a common occurrence in some areas. Although they may be relatively common, they do not occur regularly and are therefore *arhythmic,* that is, without rhythm. Their effects upon the community can be considerable. Other catastrophic events are also arhythmic. Landslides, avalanches, earthquakes or intervention by man may have a decisive effect upon any community.

QUESTIONS FOR DISCUSSION

1. How are populations related to communities?
2. How is a community a self-sufficient unit? Why is it not a true ecosystem?
3. Why do communities change?
4. What kind of evidence is available to support the idea of evolution of a community?
5. What happens to the animal populations of a community as the plant populations change?
6. What is a climax community? Why does succession occur?
7. Describe the process of primary succession as it might occur on a bare rock.
8. Why are ponds considered as transient features of a landscape? What factors might cause some ponds to undergo successional changes faster than others?

9. Briefly describe three primary causes of succession.
10. How is secondary succession different from primary succession? Give an example of secondary succession, and carry it to a climax community.
11. Under what conditions might a climax community be other than a beech-maple community?
12. How is the physical environment affected by the floral community?
13. Under what conditions is the stability of a climax community subject to change?
14. Describe how the motion and relative positions of the earth and sun cause the progression of seasons on the earth.
15. Describe how the seasons affect the biological complexion of a community.
16. What climatic factor might be responsible for a grassland climax community in the same lattitude with a forest climax?
17. Explain why man can be described ecologically as a forest-edge species.
18. What is an ecotone? Why does it occur?
19. Explain how fires can "protect" a community.
20. What particular hardships are encountered by organisms living in the intertidal zone? How are these hardships overcome?
21. Does man possess a circadian rhythm? Explain.

INVESTIGATIONS AND ACTIVITIES

1. Write a brief review of Mark Oberle's article, Forest Fires: Suppression Policy Has Its Ecological Drawbacks (*Science*, 165(3893): 568-571.)
2. Using bean plants, attempt to verify some of Wilhelm Pfeffer's observations of circadian rhythm. Write up your results in standard report form.
3. Design and carry out an experiment to determine the pattern of circadian rhythm in an insect or small mammal. Write up your findings in a standard report form. What advantage might the organism derive from such rhythmic behavior?

4. The importance of pollen analysis has been stressed in the chapter. Using the Research Problems in Biology, Series 4, No. 143, as a guide, attempt to see if the pollen content in modern animal droppings reflect the local environment.
5. Read the two articles in *Scientific American* for September 1962 describing the oceanic life of the Antarctic (pp 186-210) and the terrestrial life of the Antarctic (pp 212-230). In a brief report, compare the two communities and account for differences in terms of the environmental factors.
6. In his statement quoted on page 94, Herbert Ross pointed out that environmental tolerances of various species are important in determining the successional pattern of a community. Using seeds of one species from each of the following groups of plants, demonstrate the variations in tolerances to light and temperature. Place at least 10 seeds of each species under each of the four conditions. (c.f. BSCS Green Version, pp 254-255.)

Group Number

1.	2	3	4
tomato	radish	lettuce	vetch
pepper	tobacco	ev. primrose	larkspur
carrot	corn	onion	columbine
sunflower	flax	mullein	celery
cotton	bindweed	fireweed	

Condition

1. Continuous light & cold
2. Continuous light & warm
3. Continuous dark & cold
4. Continuous dark & warm
5. Variable light & temperature

RECOMMENDED READING

Beck, Stanley D. 1963. *Animal Photoperiodism.* Holt, Rinehart & Winston, Inc., New York.

Brown, F. A. 1959. *Living Clocks.* Science. 130: 1535-1544.

____1959. *The Rhythmic Nature of Animals and Plants.* Amer. Sci. 47:147-168.

____1962. *Biological Clocks.* BSCS Pamphlet Series, No. 2, D. C. Heath & Co., Boston.

Carpenter, J. R. 1934. *Fluctuations in Biotic Communities. I. Prairie-forest Ecotone of Central Illinois.* Ecology. 16:203-212.

Curtis, J. T. 1955. *A Prairie Continuum in Wisconsin.* Ecology. 36: 558-566.

Elton, C. S. 1958. *The Ecology of Invasions by Animals and Plants.* John Wiley & Sons, Inc., New York.

Haag, W. G. 1962. *The Bering Strait Land Bridge.* Sci. Amer. 206(1): 112-120

Harker, J. E. 1958. *Diurnal Rhythms in the Animal Kingdom.* Biol. Rev. 33: 1-52.

Hartesveldt, R. J. 1964. *Fire Ecology of the Giant Sequoias.* Natur. Hist. 73(12): 12-19.

Hendricks, Sterling B. 1963. *Metabolic Control of Timing.* Science. 141: 21-27.

Knight, Clifford B. 1965. *Basic Concepts of Ecology.* Macmillan Co., New York.

Livingston, D. A. 1955. *Some Pollen Profiles from Arctic Alaska.* Ecology. 36:587-600.

Pruitt, W. O. 1960. *Animals in the Snow.* Sci. Amer. 202(1): 60-68.

Ross, H. H. 1962. *A Synthesis of Evolutionary Theory.* Prentice-Hall, Englewood Cliffs, N. J.

The dominants in a mixed-grass prairie community may include several species of grasses such as little bluestem and gramagrasses. But the community will also include a variety of other plants. although in much smaller numbers.

6

"Every kind of organism, from a manatee to a muskox, from a palm to a pine, is the visible expression of adaptation to a particular set of environmental conditions."
Hylander*

Community Structure

How would you find out about the community in which you live? Although there are undoubtedly many ways, you might use the telephone directory. Looking through the "white" pages would give you only the names of the people living in the community – very little information when it comes to knowing about what goes on there. The yellow pages on the other hand, would be very helpful in finding out about the activities of that community.

In the same manner, if we are to find out about a natural community, we should concern ourselves with more than just what organisms occur there. We have already seen how no organism is really alone in the community. He is part of a population which is subject to outside pressures and internal activities. He is part of a community that is in a dynamic equilibrium, with each contributing member having a small, but important and necessary part to play. Just what these specific parts are, we shall explore further in this chapter.

COMMUNITY DOMINANTS

In natural communities, there are an overwhelming number of species. If you were to take an inventory of all of the species of plants and animals in a square mile of any natural community, you would soon find the task extremely laborious, if not

*Hylander, Clarence J. 1966. *Wildlife Communities from the Tundra to the Tropics in North America.* Houghton Mifflin Co., Boston.

impossible. At first, it might appear relatively simple. In a square mile of grassland, you would quickly point out five or six species of grasses, a dozen or so other herbaceous plants, perhaps two or three woody shrubs, and even some species of trees in a ravine. For consumers, a quick survey might show some deer, pronghorn, a couple of species of rabbits, probably several species of birds and field mice, and numerous insects. But the list would be far from complete.

There would still be a wide variety of soil bacteria and fungi, (fuń-ji). Other soil organisms would include several kinds of worms, other insects, and a myriad of other microorganisms. In addition, many of these organisms, as well as the more obvious ones listed above, carry other organisms around with them as parasites, both internally and externally. By now, you should begin to understand why a complete list of the organisms of a community might well be impossible to compile.

Fortunately for the ecologist, however, a complete list is not necessary. He can begin to understand how a community functions by studying the *dominant* organisms of a community. It is a general postulate in ecology that communities contain only a few species that are relatively common, and a comparatively large number of species that are rare at any given time. Naturally, the complexion of a community will change through the seasons, but at any given time, this postulate will hold true.

Figure 6-1 illustrates the relationship between the number of species present in a community and their relative abundance. Note that there will be a few species with a large number of individuals, as predicted by the postulate. In the same manner, there will be a large number of species for which only a few individuals are representative. This point may be further illustrated by a specific instance (Table 6-1).

Even though the relatively rare species do not make up a very large portion of a community, this is not to say that they are unimportant. In what way are they important? Why would it not be just as good to have a trophic level composed of only one kind of organism? What advantage exists, if any, in a community with complex food webs over a relatively simple community with simple food chains?

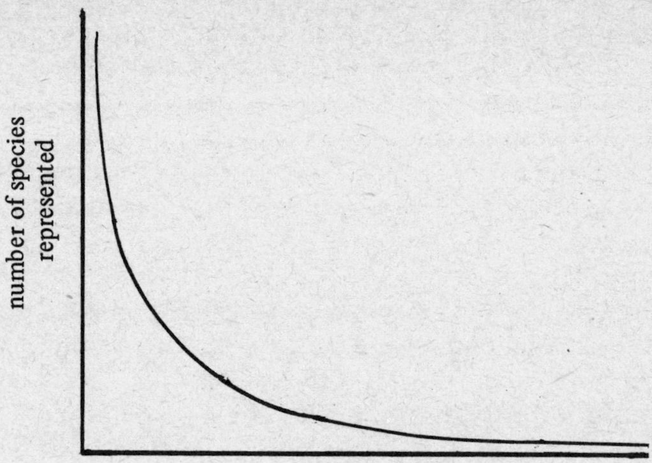

Fig. 6-1. The relationship between the number of species in a community and their relative abundances.

Table 6-1. The percentage cover of grasses and forbs in a Texas Grassland. (From Odum, 1959, after Dyksterhuis, 1946.)

Species	Percentage Cover
little bluestem	64.66
side-oats grama	8.11
Indiangrass	5.12
dropseed	2.10
hairy grama	2.50
other grasses	2.90
perennial forbs (assorted)	8.21
annual forbs (assorted)	2.30

There are no specific, clear-cut answers to these questions. The most evident answers, however, lie in the survival of the community itself. A community with a wide variety of organisms, even though some are comparatively rare, is more protected against environmental interruption than a more simple community. The community gains a certain degree of stability in the diversity of organisms which comprise it. With great diversity, the community possesses greater adaptability to changing conditions. Changes may be short term interruptions by man or even long term changes induced by alterations in the climate of the area. Whatever the cause of the altered condition, greater diversity in the community means that there will be a greater chance that some species will likely be able to tolerate the new conditions. Thus, they will be able to take over and ensure the survival of the community.

For example, a Rocky Mountain forest in which the dominant tree species may be fir and spruce, may also contain an occasional ponderosa pine, aspen and widely distributed, but far from dominant, grasses. If that forest were to suffer interruption by winter snowslides, fire, disease, or some other catastrophe, the changed conditions may not favor the immediate repopulation of the area by the original dominant trees. Under such conditions, the aspen might be quick to take over to rebuild the community, and eventually yield, under successional changes to the original forest dominants. If conditions were especially dry, the pines and grasses might serve to repopulate the stricken area. These relatively rare species of the community are thus able to protect the area from total destruction.

Species diversity can have its practical applications as well. One of the major problems facing our society today is that of pollution. It is indeed fortunate for man that his environment contains organisms that can tolerate his effluence. Without diversity among the inhabitants of the natural world, the potential for adapting to the modern world might have been absent. The consequences of such a possibility would spell certain disaster in any language.

On the agricultural front, species diversity should be kept well in mind also. There is a strong tendency for farmers to concentrate on single crops (a practice often referred to as *monoculture*). Even within single crops, there are many varieties which offer resistance to drought or certain insects and diseases. Some of these same varieties offer advantages in high yields as well. Even though high yields mean increased profits, it is still risky business to reduce diversity for the sake of another dollar in the pocket. The sudden influx of an unexpected disease to which that species or variety is not resistant might mean the difference between disaster for a farmer without diversity, and success for a well diversified farmer. Any financially successful investor will quickly point out the advantages of diversity of financial assets.

ADAPTATION

Even without a course in ecology, you undoubtedly already associate various plants and animals with a particular kind of environment. The cactus brings to mind the desert; evergreens, the mountains. The cypress swamp with its exotic birds and alligators also is characteristic in its associations. While the prairies may no longer sport large herds of bison, these massive animals are still stereotyped as representative of the Great Plains. There is also probably a vague understanding among most people that the physical environment is somehow responsible for these associations. What factors govern the distribution of life? Why can some plants and animals live where others cannot? The answers to these questions lie generally in the realm of *adaptation*. Plants and animals are somehow adapted to the specific factors of the physical environment. But to understand the phenomenon of adaptation, we must be aware of what effects the specific features of the environment have on the essential life activities.

Every organism must somehow cope with four aspects of its environment: soil, air, water and sunlight. In a natural environment, these features may be combined in an almost endless number of habitats, each having its characteristic features. Not all organisms will be affected by these features of the environment to the same degree or even in the same way. Perhaps no single feature will determine whether a plant or animal will live in a particular habitat. More likely, the specific combination of physical conditions will be more important than any single factor.

Soil. The soil is a complex environmental feature. It may be rocky or relatively smooth-textured. Its particular chemistry may be high in certain minerals, low in others, or be somewhat acid or basic, depending upon a variety of factors. Chemical substances, particularly organic acids which are added to the soil by the plants and animals which live in the area may help to decompose the parent rock material and at the same time modify the particular soil chemistry. As water percolates into deeper layers of the soil, it may carry with it many of the minerals and other chemicals. This *leaching* process in turn may affect the life forms living in and on the soil by making some substances unavailable.

Tundra soils are found in the northern parts of North America and in similar localities throughout the world. The overlying vegetation is composed of mosses, lichens, dwarfed shrubs and relatively small herbs. Being rather flat, drainage is poor. Permafrost occurs at relatively shallow depths, preventing water from leaching into the lower soil layers. The result is a rather peaty surface layer. Further south, where moisture is more abundant, permafrost is absent or at least much deeper. Cool temperatures prevent the rapid decay of organic debris, allowing chemicals to be leached away from the surface layers. The accumulation of the organic debris produces an acidic soil condition favoring a rich fungus population, but restricting many soil fauna. The degree of acidity of a soil may also influence the rate at which other plants are able to absorb soil nutrients.

The underlying limestone parent rock throughout most of the midwest United States is responsible for the basic nature of the soil. Prairie soils are characterized by high humus content since the grass plants are relatively short-lived and their organic materials are added to the soil in large amounts. Decay is rapid, releasing these materials to the soil. Mineralization, on the other hand, is relatively slow in these soils. In the tropics, decay is very rapid and humus is not allowed to accumulate. Farm land in such areas is therefore readily depleted.

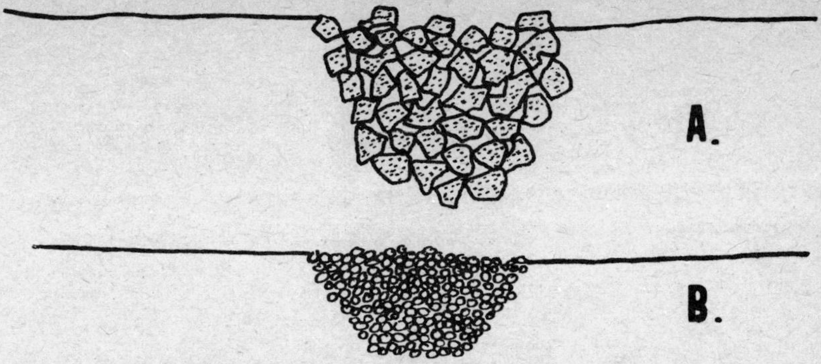

Fig. 6-2. A comparison between a coarse-textured soil (A) and a fine-textured soil. Note how the différences in the pore size are caused by the differences in the particle size. (Shaded areas represent particles.)

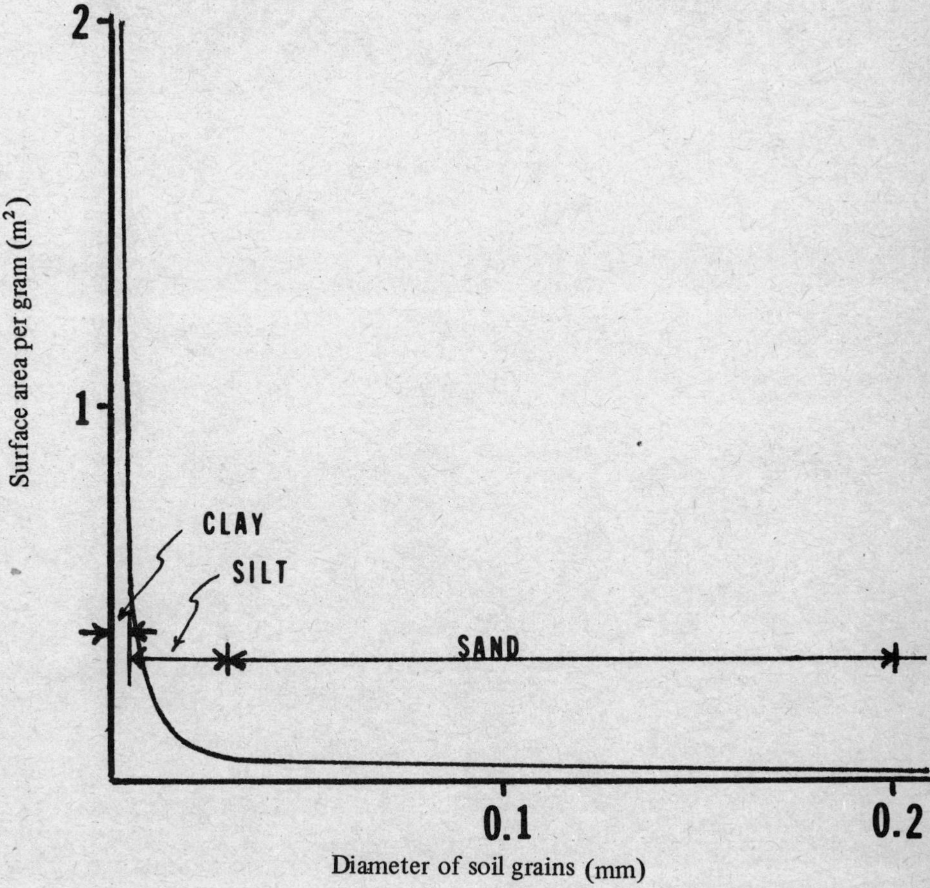

Fig. 6-3. The relationship between particle size and the surface area in different types of soil. Smaller grain diameter greatly increases the surface area upon which water can be adsorbed.

Water balance in pine trees is aided by reduced leaf surface area. Such an adaptation reduces the amount of water that evaporates from the plant.

Soils also vary considerably in their ability to hold moisture. Coarse soils tend to be drier than fine soils because there is less surface area exposed for water *adsorption* (Fig. 6-2 and 6-3). Because less water is adsorbed onto soil particles, more water is available to the plants and animals which inhabit the soil. In addition, coarse soils allow plants and animals to penetrate the substrate more easily and thereby aid in aerating the soil and mixing its various components.

On the other hand, fine-textured soils do not allow water to pass through to deeper ground layers as rapidly where it becomes unavailable. Consequently, more water is retained in the upper soil layers where it is available for a longer period of time following a rain. Evaporation is also retarded in fine-textured soils because of the small soil pore size. The small pore size, however, favors the rise of water from underlying water tables due to increased *capillarity*. Farmers are well aware of this capillary action in packed soils. To help retain moisture, they are quick to break up the ground following a harvest when warm weather would otherwise quickly rob the soil of much of the remaining moisture. By breaking up the ground, particles are not so close together and capillarity is retarded.

As you might expect, the wide variability among soils is reflected in the types of animals and plants which are found in association with them. Where water is quickly drained to lower levels and not available at the surface, plants may adapt by sending long taproots into these lower strata. Burrowing animals will be more common in areas where penetration of the soil is easy. On the other hand, where soil is very fine, it is usually compact, restricting the activity of burrowing animals. High salt concentrations in some soils may prevent the growth of certain plants and animals to only the most tolerant species. Thus, the soil offers conditions to which plants and animals must adapt in order to survive.

Air. Perhaps the single most significant aspect of the air to which organisms must adapt is that of temperature. Among plants, the reproductive activities are particularly subject to limitations imposed by temperature extremes. Late frosts are commonly responsible for the destruction of flowers or young fruits especially in fringe areas of a plant's distribution. In such

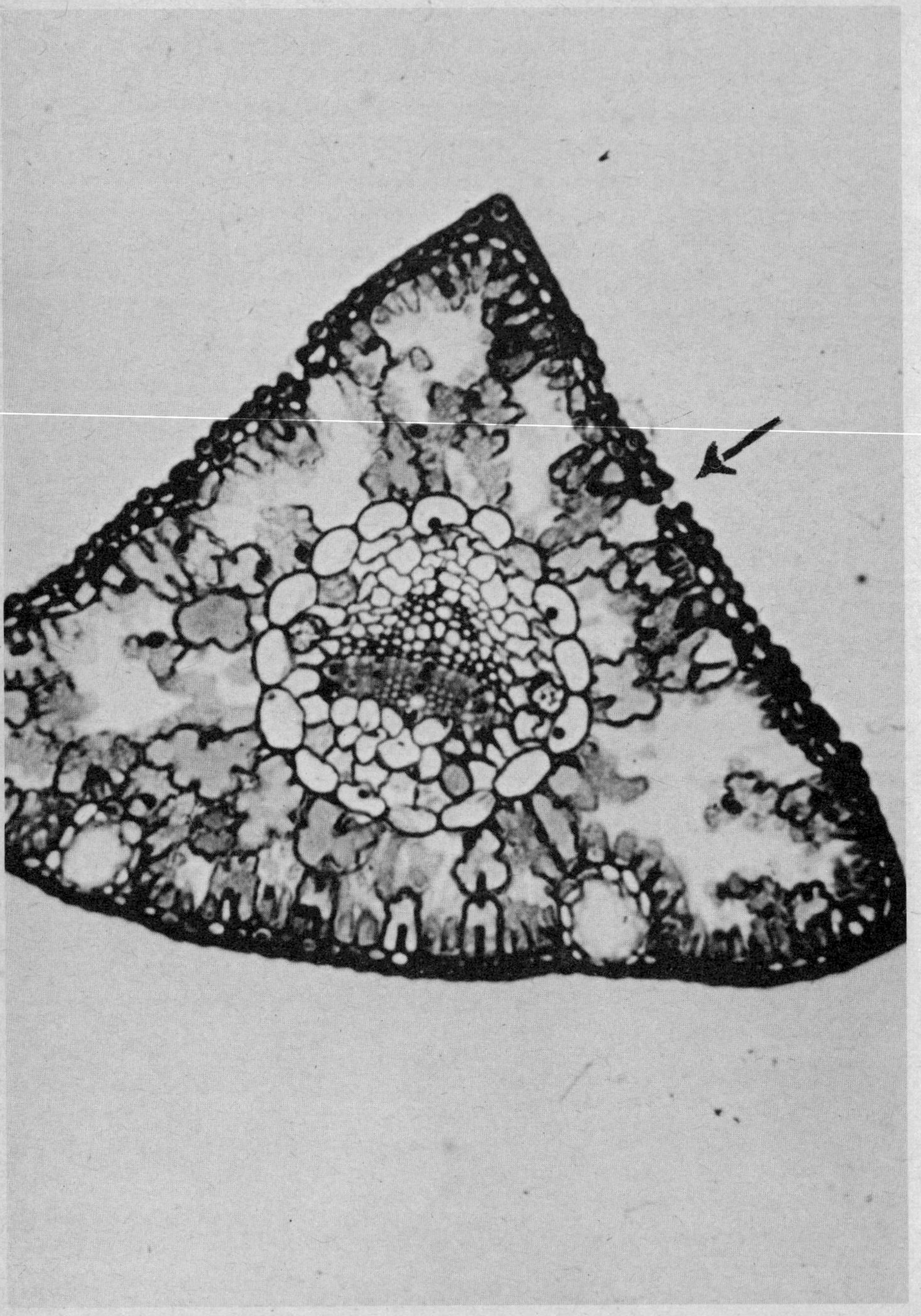

Water evaporation from the pine leaf as shown here in cross-section (X100) is also hindered by the recessed stomate (arrow).

areas, environmental factors already limit the further spread of the plant resulting in fringe area plants being particularly vulnerable. Many plants may be grown as ornamentals in areas where they cannot produce fruits due to temperature limitations.

Where seasonal climates are responsible for especially low temperatures, the ground is often frozen. This naturally reduces the amount of water available to plants. Low temperatures also will reduce the process of photosynthesis. Normal water balance within plants must be maintained, however. One way this is accomplished is to reduce transpirational water loss.

Water balance in evergreen trees (such as pines, spruce, firs, junipers, etc.) is maintained during such periods of stress by specialized leaves that offer reduced surface areas for evaporation. Needles or scales replace the leaves found on other (*deciduous*) types of plants. In addition, a heavy waxy coating further retards water loss. Photosynthesis can still occur, during warm periods, to a limited extent in these trees since their leaves are retained. The evergreens are also adapted to the relatively dry conditions that occur in the summer months. The needles and scales again offer the advantage of a reduced transpirational surface. The fact that the *stomata* (small openings in the leaves through which water vapor and atmospheric gases may pass) are recessed into depressions in the surface of the needles (Fig. 6-4) also aids in retaining moisture. The depression creates a *microenvironment* of relatively high humidity, out of the wind, thus reducing evaporation from the leaf.

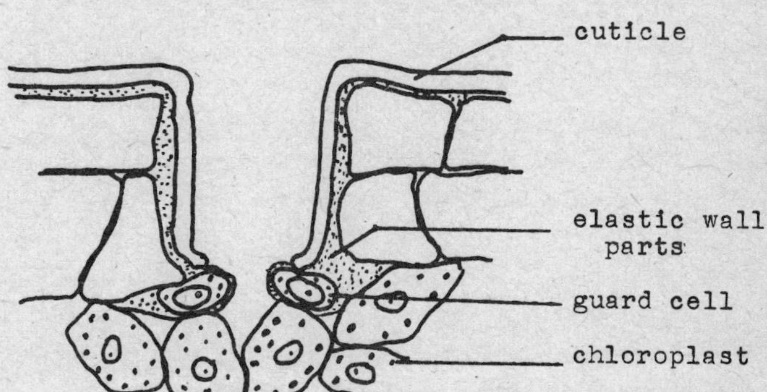

Fig. 6-4. Some evergreen needles have recessed stomata as an adaptation to xeric conditions. In addition, most leaves possess a waxy surface coating, the cuticle, to further retard water loss.

Ordinarily, high environmental temperatures are only a problem to plants when the water supply is limited. When this is the case, adaptation to hot, dry conditions may take the form of reduced leaf size, reduced number of stomata, specialized reproductive processes, and especially extensive root systems.

Animals also exhibit characteristic adaptations in response to temperature. Warm-blooded animals (*homeotherms*) are able to maintain a constant body temperature, even in cold weather. Thus, they can remain active during the winter months. They are further adapted to cold conditions by the presence of fur or feathers. However, winter poses an important problem besides low temperatures. There is normally a reduced food supply. Where winters present extremely cold conditions or make food difficult to obtain, many animals will *migrate* to more favorable areas. Elk and deer, foraging at high altitudes during the summer, will move to lower elevations during the winter. Ducks and geese follow their familiar flyways from their summer breeding brounds in Canada and northern United States to winter feeding grounds in the south around various parts of the Gulf of Mexico. Other migratory habits may be less familiar, but examples are common throughout the animal kingdom.

A second method of adapting to winter conditions among homeotherms involves assuming a dormant condition. Many mammals, such as the woodchuck go into *hibernation.* In true hibernation, the body temperature is reduced to the 40-50°F range; heartbeat and breathing rate are also sharply reduced. With such lowered bodily activity, little reserve food is used, and the animal is able to keep alive during the long winter when food is not available and temperatures would normally reduce the survival of these organisms.

Although commonly referred to as hibernation, a third adaptation to cold weather is simply *dormancy*. Bears and chipmunks become dormant, or temporarily inactive, and are thereby also able to conserve food reserves and body fat. The dormant condition is often interrupted by periods of activity. When the weather warms up temporarily, the animals may move about and forage for food.

Poikilotherms (cold-blooded animals) such as reptiles and amphibians, are not able to control their body temperatures by internal mechanisms as do those animals which are warm-blooded. When the environmental temperature falls below approximately 40°F, most of these organisms cease to be active. Winters thus pose a special problem to which these animals have adapted by retreating into burrows or burying themselves in the mud at the bottom of a lake or pond.

Animals also have adaptations which enable them to survive in hot environments. Warm-blooded animals are constantly in the process of removing heat by removing water from their bodies. Some animals may sweat. The evaporation of water is an effective coolant because water is capable of absorbing large quantities of heat. When it evaporates, it takes this energy with it, effectively cooling the body. Other animals may breathe heavily and quickly. The panting of dogs on hot days is a familiar example where large amounts of moisture may be removed to aid in cooling the body. Poikilotherms usually cannot endure the high temperatures that mammals and birds can. Consequently, even though reptiles are relatively common in desert areas, and the association of reptiles with desert areas is often made, their behavior belies their susceptibility to the high temperatures normally encountered there. Activity is usually restricted to mornings and evenings. The warmest hours of the day will be spent in the shade of rocks and vegetation or in burrows with activity being limited to quick forays for food. Some poikilotherms may survive extremely hot periods by *estivating*. This process is a special dormancy somewhat similar to that of hibernation.

Water. To aquatic organisms, procurement and conservation of water present no serious problems to survival. However, some aquatic habitats are subject to dry spells. During hot summers, ponds and small lakes may completely dry up. Roadside ditches, occasionally holding water for several weeks, are ideal for a large variety of microorganisms. But such habitats are usually only temporary at best. Most aquatic microorganisms have adapted to intermittent dry spells by developing a dormant form capable of withstanding long periods of dryness. The profusion of organisms that develops when a handful of dry grass from any roadside ditch is placed in sterilized water for a few days is ample proof of the ability of these organisms to withstand periods of drought.

Aquatic organisms do face one serious problem related to the maintenance of a proper water balance within their bodies. Under normal conditions, water is constantly passing into and out of the cells which make up the living bodies of all organisms. But the membrane which surrounds each cell is capable of restricting the passage of some other particles through it and permitting others to *permeate* (or pass through). In describing this situation, we say that the cell membrane is *selectively permeable.*

With no membrane, we would find that particles in a solution tend to move around and distribute themselves evenly throughout the solution. If they are highly concentrated in one area, their motions (diffusion) will eventually cause them to move to areas of lower concentration until they are uniformly distributed in concentration throughout the solution.

Suppose, however, that a solution which contains a high concentration of salt particles is separated by a selectively permeable membrane from another solution containing a lower salt concentration. Suppose, further, that the membrane is not permeable to salt. Under such conditions, the salt will not be permitted to diffuse through the membrane. The concentrations on either side of the membrane will still tend to be equalized, however, because the movement of water through the membrane is not restricted. Water particles will move from the area where they are high in concentration (and the salt is low in concentration) to the area on the other side of the membrane where water is low in concentration (and the salt is high). This diffusion of water through a selectively permeable membrane from a region of high concentration to a region of low concentration is known to biologists as *osmosis.*

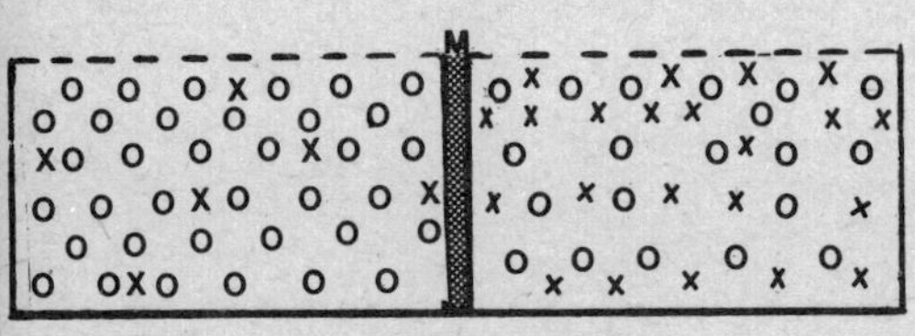

O = water
X = salt
M = selectively permeable membrane

direction of osmosis
(net water movement)

In the oceans, the surrounding water contains a high concentration of salt (and consequently, a relatively low water concentration) with respect to the cells of the organisms which live there. Thus, the organisms find themselves in an environment which favors the loss of water to the environment by osmosis. Freshwater organisms face exactly the opposite situation. Both types of organisms must be adapted to their particular situation in order to maintain a proper water balance within their bodies. Losing too much water will cause their cells to dehydrate, even when the animal or plant is surrounded by water. Taking in too much water will cause the cells to swell and eventually rupture, resulting in their death.

Terrestrial plants and animals must also possess special adaptations to cope with life on land. The availability of water to these organisms may vary from great scarcity as in a desert to unlimited quantities as in a marsh or rainforest. Where the air is moist, organisms will be able to conserve moisture because evaporation of water from body surfaces will be somewhat inhibited. Such areas are often characterized by an abundance of *epiphytes.* These plants are attached to trees for support only and are able to obtain needed moisture directly from the air. Epiphytes include a wide variety of plant forms. Most obvious are several species of orchids and the Spanish "moss" common in the southeastern United States. Even more common, but often unnoticed are many species of algae, mosses, liverworts, lichens and some fungi and ferns which grow on the bark of trees. Tropical regions, with higher and less variable humidity, will have a greater variety and abundance of epiphytes than the more temperate United States.

In drier regions, loss of moisture by evaporation presents a problem which is overcome by the development of special adaptive mechanisms. Reduction in leaf size has already been noted among the evergreens as a method for conserving moisture. Some plants, such as the Mormon tea, possess only small, scale-like leaves, the leaf size being reduced to the point that photosynthesis is carried on entirely in the green stems of the plant. The number and distribution of stomata on the leaf surfaces and the development of thick, impervious waxy coatings also control water loss. Plants growing in relatively dry areas may have very few stomata, these being restricted to the lower leaf surfaces. Such an arrangement will reduce transpiration to a minimum.

Organisms are often broadly classified into three basic types, based upon the type of area in which they are found. *Xerophytes* are plants of dry (xeric) areas. Representatives of this group include the cactus, creosote bush, mesquite and the Mormon tea, to mention only a few. Plants which grow in water or marsh areas represent the opposite extreme. These are known as *hydrophytes* and include cattails, tapegrass, water lilies and water hyacinths. Between these two extremes lies the broad range of the *mesophytes.* Plants of the mesic (medium) areas would include most trees, shrubs and herbs with which you are familiar.

Naturally, there is no distinct dividing line between any of these three categories. Consequently, there is a broad continuum represented by forms which exhibit gradations from one extreme to another. Furthermore, classifying plants into these groups is usually quite relative. Oak trees, for example, would be regarded as generally mesic plants. Compared to beech or maple trees, however, they would be considered as relatively xeric species. Some of the fundamental causes behind succession become more apparent in light of the adaptations to moisture possessed by various organisms.

Conservation of water among animals also takes many forms. In humid areas, or where a water supply is reasonably abundant, no special adaptations are required. However, where water is scarce, the animal is faced with a problem of survival. Two types of adaptations have met the challenge. First, most animals in dry environments have developed a means for conserving water, enabling them to go long periods without water. Reptiles are perhaps the most common examples. The impervious scaly coating is quite well adapted to restricting water loss. In addition, these organisms, as well as others, may spend the hot periods of the day when humidity is especially low, in burrows where a higher humidity can be maintained. Other animals, such as the camel, may be able to actually store water which can be used over a long period of time.

The second method by which animals may adapt to water scarcity is exemplified by a small number of rodents such as the kangaroo rat, and a few reptiles such as the horned "toad." These animals are able to manufacture their own water in sufficient quantities by the breakdown of sugars during respiration. Most animals and plants do this also, but the amounts of water produced are not nearly sufficient to sustain them. Such an adaptive mechanism is truly a spectacular achievement for the processes of nature.

Light. As do many other physical factors, light may exhibit seasonal variations. It has already been pointed out in Chapter 3 that the intensity of the sun's rays is dependent upon the angle with which they strike the earth. Furthermore, the angle is dependent upon the particular season of the year, or more precisely, the position of the earth in its orbit. Where a plant grows is usually related, among other things, to its ability to

In the lower elevations of the Rocky Mountains (6000–9000 feet above sea level), north-facing slopes will be cooler and will therefore retain moisture and allow the growth of trees, especially the Douglasfir. South-facing slopes are exposed to more direct, drying solar radiation, and few, if any, trees will be able to tolerate the lack of moisture.

tolerate conditions of direct sunlight or limited sunlight. Sunlight, then, can be a limiting factor with two extremes just as moisture, temperature or other physical factors can be. Particular plants will be adapted to certain light intensity situations in the number of *chloroplasts* per cell, size of leaves, and the arrangement of the leaves on the plant.

Animals, too, are subject to adaptative mechanisms with respect to variations in light intensities. Although perhaps less noticeable in their distribution than plants, animal distribution may also be at least partially in response to particular light intensities. Nocturnal animals have special adaptations enabling them to see well in dim or dark situations. Animals active by day depend upon light to find food and see their enemies.

Seasonal variations in the relative lengths of daylight and darkness (photoperiod) are also felt among the biological world. The flowering process in many plants and seasonal migrations of a variety of animals have already been mentioned.

Many other physical factors of the environment are undoubtedly responsible for the adaptation of organisms to particular niches and habitats. Most of these situations are unique and may actually be considered as caused directly or indirectly by combinations of the four basic elements of the environment mentioned above.

Topography, or terrain of the land may influence the types of life found in particular situations. A typical example may be found in the Rocky Mountains. Figure 6-5 illustrates the differences in vegetation that may be found between north-facing and south-facing slopes. North-facing slopes receive little, if any, direct solar radiation. This influences the temperature extremes that will be found on such slopes. Temperature directly influences the rate of evaporation of water from the soil. Consequently, a cooler, more moist environment on such slopes will be a sharp contrast to south-facing slopes where direct sunlight ultimately is responsible for the existence of more xeric conditions. The result will be obvious differences between the two slopes with respect to the vegetation and the animals which are in association with particular vegetational communities. Douglasfir will be the dominant tree on north-facing slopes; a less-dense forest of ponderosa pine and Gamble's oak predominate on the south-facing slopes.

Fig. 6-5. Differences in vegetation as influenced by topography. A, an eastern United States forest; B, central Colorado in the Rocky Mountains.

Fig. 6-6. The "flagging effect" produced when wind blows constantly from the same direction at relatively high velocities.

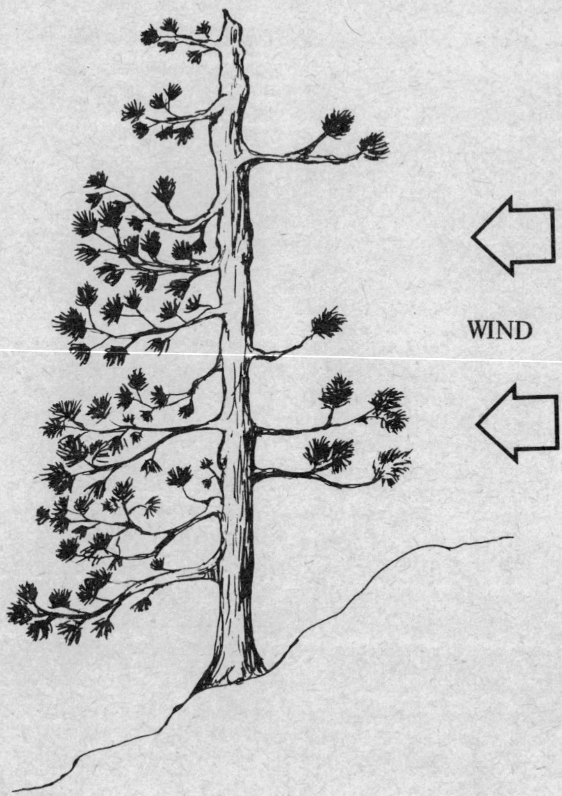

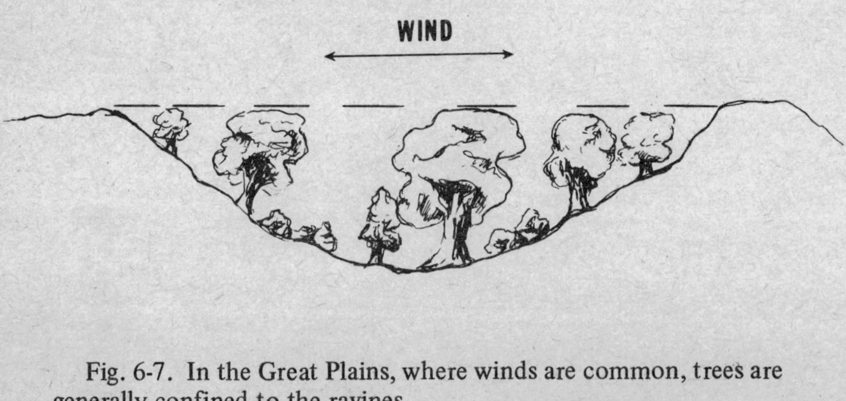

Fig. 6-7. In the Great Plains, where winds are common, trees are generally confined to the ravines.

When wind blows constantly from one direction at high velocities such as on mountain tops, the few hardy trees may exhibit flagging, with branches only growing on the lee side of the tree.

Wind may also be an important factor. But this factor may also be a result of combinations of other physical factors. Solar radiation and its heating effects will particularly influence direction and intensity of wind. This is especially noticeable in the land and sea breezes of the coasts. Other factors also are responsible, of course, but it is important to not divorce or isolate one factor from others. The effects of wind on the biological world is readily demonstrated in the "flagging effect" (Fig. 6-6) found where trees are constantly subjected to high velocities of wind from the same direction. Wind may also restrict the growth of trees to small, stunted forms or prevent their spread into areas surrounding heavily treed ravines (Fig. 6-7). Wind, by carrying sand particles, may also alter the topography of an area over a long period of time by carving out the softer rocks, leaving those which are more resistant.

STRATIFICATION

The structure of a community is not limited to merely the kinds and numbers of organisms living in it. That each organism occupies a rather specific niche in the community has already been pointed out. However, to perform the required activities demanded by his "profession," the organism usually occupies a rather specific habitat as well. Thus, careful examination of a community should reveal a characteristic *stratification* for each type of community. Generally speaking, two basic strata can be identified. There is the upper zone, exposed to direct sunlight, and responsible for most of the photosynthetic activity of the community. Dominant organisms in the upper stratum are autotrophic. Beneath this zone is the consumer zone with the dominant organisms being heterotrophic.

The upper zone is usually further subdivided into several rather distinct layers (Fig. 6-8). In a forest community, for example, the uppermost story is made up of sun-tolerant trees. These trees form a *canopy* over the lower parts of the community which may vary considerably in its openness. Beneath the canopy there will often be several more or less distinct layers. A layer near the canopy, but never exceeding it, may exist. Such trees will be shade tolerant and their density is

Fig. 6-8. Typical stratification in a deciduous forest.

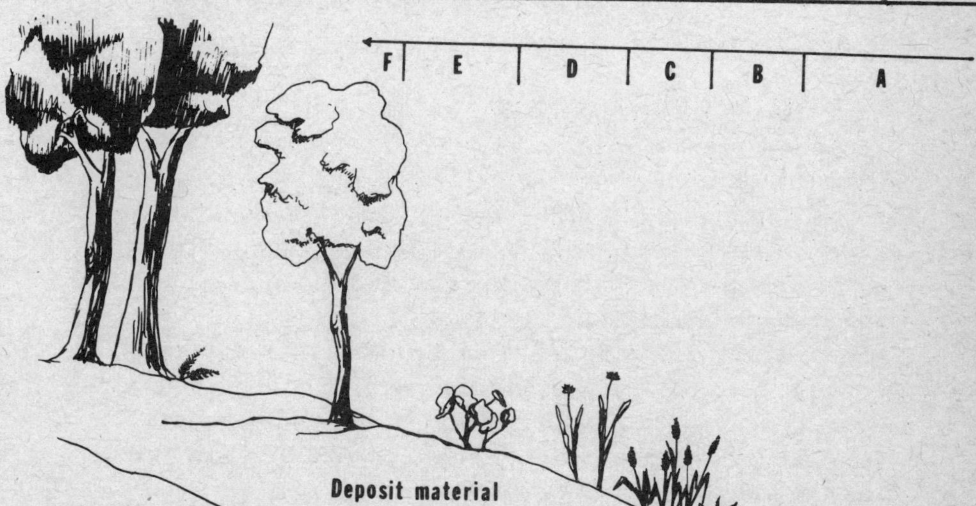

Fig. 6-9. Stratification of freshwater vegetation from the completely submerged type (A) through the floating type (B), emergent hydrophytes (C), sedges and grasses (D), shrubs and pioneer trees (E), and finally, the climax forest (F). (From Knight, 1965.)

usually considerably less than that of the overstory. A third layer composed of shorter shade-tolerant trees may also be present. Included in this layer, extending from approximately four feet to the bottom of the canopy layer, will be trees which will extend into the canopy when mature and either increase its density or replace it completely in the process of succession.

Closer to the ground there will be a stratum composed of shrubs and small tree seedlings which are, of necessity, shade tolerant species. And below this is the herbaceous stratum, the last of the obvious strata. There will be others, however, beneath the surface of the ground which can be distinctly identified. Grassland communities will also be stratified. However, the thickness of each layer as well as the community as a whole will be considerably less than that of a forest community. Separating and identifying these strata would be more difficult.

Aquatic communities exhibit stratification as well as terrestrial communities. In lakes and ponds, horizontal zonation is characteristic of the successional stages outlined in the previous chapter. The appearance of such communities, therefore, will be a blend of both zonation and stratification (Fig. 6-9). Emergent hydrophytes such as cattails and rushes will mark much of the shallow periphery of a pond. Further out, floating plants rooted in the bottom mud will shade lower levels and restrict most submerged vegetation. Plants in this zone might be water lilies and water hyacinth. However, the water depth soon limits these plants and the upper story in open water is represented by phytoplankton. Depth will eventually limit these producers also for two principal reasons. First, sunlight cannot penetrate to great depths with enough efficiency for the photosynthetic process to continue. Second, the multiplication of phytoplankton under certain conditions may become so rapid that their great numbers prevent the sun from reaching even normal depths. Submerged rooted vegetation such as water weed, tapegrass and water crowfoot is also restricted to relatively shallow waters for the same reasons.

Faunal stratification also occurs in most communities. The distribution of flora into a variety of strata will increase the number of habitats as well as the number of niches available for occupancy. Competition may thus be reduced and a larger number and greater variety of organisms are then able to make more efficient use of the solar energy reaching an area. You might expect, therefore, that communities with the greatest number of strata will exhibit greater species diversity and a larger total number of organisms. Such is actually the case. Tropical rainforests are among the most richly diversified communities in the world. There is a more than sufficient supply of moisture throughout the year. The directness of solar radiation results in a greater amount of energy per unit area striking the earth in tropical areas than in higher latitudes. Consequently, such ideal conditions have fostered a community where time has allowed extremely efficient exploration of every available niche.

Animals, by their very nature, are mobile and therefore not necessarily confined to certain strata. The daily activities of birds may bring them into several strata in search of food or flight from predators. Insects may require more than one stratum to complete their life cycles. Other organisms may use a certain stratum for feeding, another for breeding, and still another for sleeping. Thus, although several strata can be identified and described, they remain linked together and interdependent. For this reason, strata cannot be considered as separate communities. Rather, they are subdivisions of a single community.

INTERSPECIFIC ACTIVITIES

Before we leave the subject of community structure, perhaps some mention should be made of the manner in which various species interact. These interactions may be quite specific and easy to define. On the other hand, the relationships may be extremely vague and are often overlooked. Nevertheless, it is through the interaction between various species that each species is able to fulfill its role in the community.

The term *symbiosis* has been used by many ecologists in many different ways. As a result, much confusion has resulted from its use. It is derived from the Greek, meaning to "live together." Since each of the forms of interaction described below (except for neutralism) requires some manner of interaction and hence "living together," the term symbiosis is used here as a blanket term describing a general relationship between two organisms within a community. This is generally the accepted sense of the term.

Species interaction may take many forms. However, for the sake of simplicity, let us consider six basic forms of interaction. There are: (1) *neutralism,* in which the relationship is such that neither population has an effect upon the other; (2) *competition,* in which there is a constant struggle between both populations for food, shelter, nesting sites, or other facet of existence; (3) *mutualism,* in which two populations may benefit each other in their survival; (4) *commensalism,* in which one population receives benefits from another which is unaffected by the first population; (5) *parasitism,* in which one population is benefited at the expense of the other; and (6) *predation,* which is similar to parasitism except that it results in the inevitable and immediate death of the attacked organism.

Neutralism. This term implies that two different species are affected neither beneficially nor adversely by their association. In some instances, when the food supply becomes limited or some other aspect of the environment is altered, the neutral relationship between the two organisms may also be altered. Competition for food or shelter may become so intense that one or both species may be severely affected. Two neutral species may also alter their relationship to one another during various parts of their life cycles. In view of this, why do you think it might be desirable to know the details about the complex life cycles of . . . certain insects, for example?

Competition. Much of the discussion involving population dynamics and community balance have implied the existence of competition as a matter of fact. The objects of competition may be diverse. Between different species, however, they are generally limited to food. Among plants, there will be constant competition for sunlight, water and minerals. The most intense competition usually occurs between members of the same species, or at least closely related species. Food, mates, breeding sites, shelter, territory and even social standing in a population are common objects of competition among members of a population. The competition does not necessarily infer that physical combat takes place, although it may occur as a last resort. Most animals, at least among members of their own species, have evolved a systematic behavior which reduces the need and substitutes for actual combat. Competition, although it appears to be a negative factor in community dynamics at first, is actually quite valuable from an evolutionary standpoint. Through intense competition, whether between members of the same or different species, the survival of the hardiest and most well-adapted members of a given species is assured.

Mutualism. The community may exhibit at least two distinct forms of mutualism. Although the close association of two species may result in their mutual benefit, in some instances, this relationship is not at all required for the survival of either species. Such a situation might be described as *non-obligatory.* One of the more common examples of non-obligatory mutualism is seen in the hermit crab which uses the empty shells of marine snails, or shells of other crabs). A variety of coelenterates, sponges and other marine invertebrates often attach themselves to the shell surfaces, thus providing the crab with a protective camouflage. They, in return, receive transportation which enables the otherwise sessile organisms to obtain food more readily. An additional benefit may be derived in the case of the hermit crab. The hitch-hiker may be a particular coelenterate which benefits the crab by building up the free edges of the adopted shell. The crab thus does not need to trade "homes" periodically since he is supplied with a home that is constantly expanding to fit his needs.

Many animals, such as birds, receive some of their nourishment by feeding upon the external parasites of other animals. Such is the case in which the rhinoceros is "groomed" by a bird which eats ticks. Presumably, ticks are not the only food for this bird, and the survival of either organism is not dependent upon this relationship.

Frequently, the intimate association of two species has become so involved that neither species can survive without the other. Thus, the second form of mutualism involves an *obligatory* relationship. Undoubtedly, the most common example is that of the lichen. Often classified as a single plant, the lichen is actually an intimate symbiosis involving a fungus and an alga (Fig. 6-10). The fungus, being non-photosynthetic, is able to obtain its required food from the alga. The fungus, in return, provides the alga with a substrate upon which to grow and also helps to maintain the water balance required by the alga.

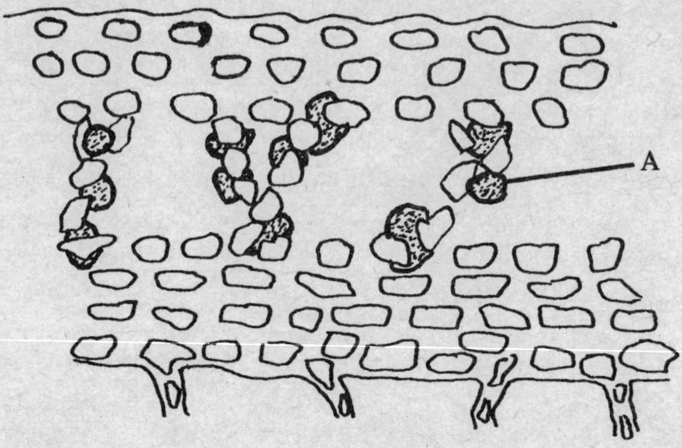

Fig. 6-10. The lichen is an intimate symbiotic relationship between an alga (A) and a fungus. The alga supplies the food for the fungus and in turn receives water and a "home" from the fungus.

The ocean presents countless examples of spectacular mutualistic relationships. Of particular interest is that between the tiny neon goby fish and the moray eel. Considered one of the most dangerous predators of the coral reef, the moray will hold absolutely still and permit the goby to swim over its body, into its mouth and even into its gills in search of parasites. Many reefs have even been identified as "cleaning stations" where fish visit periodically and routinely to have parasites removed. Specially adapted "cleaner" animals such as the goby and various shrimp provide expert parasite removal.

Countless other examples exist in varying degrees. In some instances, the intimacy is short-lived as in the mutualistic relation between a flowering plant and the insects which pollinate it. In other instances, the association is prolonged as in the lichen. Curiously, the termite, which survives on wood for its nourishment, cannot digest the needed cellulose. Small unicellular microorganisms (flagellates) live in the intestinal tract of the termite and break down this complex food material to make it available to the termite. They, in turn, receive nourishment as well as a home. Many herbivorous mammals such as the cow have the same intestinal arrangement with some bacteria, although the relationship may not be entirely obligatory as far as the cow is concerned.

Commensalism. The above example of the symbiotic relationship between a cow and the intestinal bacteria may be a borderline case between mutualism and commensalism. Often the distinctions are not clear. The barnacle that hitches a ride on a whale is a somewhat more clear-cut example of commensalism. The barnacle receives transportation, and perhaps a greater food supply thereby. The whale, however, is generally unaffected by the association. To carry it to an extreme, there are even barnacles that live on barnacles that live on whales!

The unique adaptations of many animals and plants to particular situations never ceases to be amazing. Such is the relationship between some marine fish and sea anemones. Some fish are able to live among the tentacles of the anemone, apparently immune to the blasts of stinging cells that are sometimes fatal to others.

Perhaps one of the best known commensal relationships is that between the remora and the shark. Unaffected by the hitch-hiker, the shark carries the remora to a feeding area. The remora detaches itself from the shark and swims around feeding on the bits and pieces of food dropped by the shark. When the shark moves on to another area, the remora becomes reattached. Occasionally, other fish provide the transportation for the remora. Even skin divers have been known to serve this function. There is some recent evidence, however, that may alter the accepted view of this symbiotic relationship. Some observers feel that the remora may actually pick parasites from the shark and in return receive transportation and immunity from attack. If this is true, the relationship becomes one of a mutualistic nature.

Parasitism. Some mention has already been made of the last two forms of symbiotic relationships. Although parasitism is a negative factor acting on one population, it seldom reaches disastrous proportions. There are few plants or animals that entirely escape being parasitized. Usually, the parasite possesses remarkably specialized adaptations in structure or habits which enable it to successfully extract the required materials from its host. Ideally, the host is not seriously hampered by this rela-

The mistletoe is a dread plant infection in some parts of the United States. It is parasitic on a variety of trees.

tionship since the life of the parasite depends upon the life of its host.

Equally remarkable is the evolution among the hosts of the means to at least partially resist the completely deleterious effects of the parasites. This point is best illustrated when a parasite has been recently introduced into an area occupied by a potential host. The host population, not having been in contact with the parasite previously, has not had sufficient time to evolve a defensive mechanism. As in the case of the inadvertent introduction of the sea lamprey into the Great Lakes, the results can be disastrous in many respects.

Natural control of parasites is accomplished in many instances where the parasite requires more than one host to complete its life cycle. The transmission from one host to another often represents a weak link in the life of the parasite, and its spread never reaches epidemic proportions. Some organisms are parasitic only for specific periods during their life cycles. Thus, the female ichneumon fly deposits her eggs in the body of a grub or caterpillar beneath the bark of a tree. The larvae feed upon the live grub after hatching. When the larvae eventually reach the adult stage, they become free-living.

Plants can also be parasites upon other organisms. In some cases, green plants may even be the parasite. The common mistletoe of the southern United States uses the branches of oaks and sycamores for support and additional nourishment as it sends its special roots into the tree. Deformed branches and dense mistletoe growth called "witche's brooms" are conspicuous during the winter when the leaves are no longer on the tree. Non-green plants are more commonly parasitic than green plants, however, Since they do not possess the ability to produce their own food, nourishment must come from other organisms. Fungi and bacteria represent the common forms of this type of parasitism. On trees, some bacteria may cause discoloration and wilted or deformed foliage. Other bacterial infections in plants may cause abnormal growths, often seen as galls and tumors. Bacterial growth in animals is usually more obvious in the disease it may produce. Molds, mildews and rusts, common to the agriculturalist, are examples of plants infected with fungi. The chestnut blight in the early part of this century wiped out thousands of acres of chestnut forest especially in the southern Appalachian region. Even the ever-present "athlete's foot" is an infection in which a fungus parasitizes man.

The chestnut blight is only one example of how a parasite can be brought into a new area by introducing a desirable organism from a different locality. Immigrant parasites may have no natural enemies and their rapid population growth may completely annihilate the host population.

Predation. Little really needs to be said concerning this particular relationship. The effects are also negative on the growth and survival of the *prey* population. However, such negative effects are not without their positive effects, particularly in terms of the community as a whole. Prey populations tend to become adapted by developing speed, camouflage, or other means of eluding the predator. Paralleling these developments, evolutionary adaptations among the predators aid their attempts to still obtain food. Moreover, the constant interactions provided by this relationship provide the community with a stability that is required for the survival of the community itself.

The community is thus a complex operation. It is a dynamic equilibrium with adaptations constantly keeping pace with changing conditions. Both intraspecific activities and interspecific activities are deeply involved in maintaining this delicate balance.

QUESTIONS FOR DISCUSSION

1. Compare the advantages and disadvantages of an organism being highly adapted (specialized) for its particular niche in contrast to its being a more generalized form.
2. What are the six basic forms of interspecific activities and how are they defined? Give concrete examples of each.
3. What are epiphytes? Give at least two specific examples.
4. What four basic aspects of the environment require adaptation by organisms? For each, point out a form that such an adaptation in a plant and animal might take.
5. Describe two types of adaptations in animals to living in grasslands, and give examples of animals which exhibit each.

6. What terms are used to describe a community with respect to the relative abundance of water?
7. What are the essential differences between hibernation and dormancy? For each list two animals which use that method of adapting to winter conditions.
8. What is osmosis?
9. In what significant way must marine and freshwater organisms be different in their adaptation to life in the water?
10. What serious problem does a freshwater organism face in regulating water balance?
11. What environmental factor is associated with estivation? Give some examples of organisms that estivate.
12. How can we explain the presence of large numbers of arboreal animals in tropical rainforests?
13. Many species of migratory birds of North America and Europe are being observed at higher latitudes than they were at the turn of the century. How might you account for this?
14. Define symbiosis. Which forms of interspecific activities might properly be termed symbiotic?
15. What is the difference between interspecific and intraspecific competition?

INVESTIGATIONS AND ACTIVITIES

1. Write a report on the Dutch elm disease, including the following areas in your report: its history, symptoms, vector, treatment, and epidemiology.
2. The following questions relate to soils, although not specifically covered in the text. Write a brief report in which these questions are considered. (a) How does the size of mineral particles in the soil affect its suitability for microorganisms? (b) In what ways may the acidity or alkalinity (pH) of soil water affect soil organisms? (c) Desert soils usually contain a relatively low percentage of humus and have a low density of soil microorganisms. Why are they

often very productive of crops? (d) Wild plants are often difficult to transplant. List as many reasonable explanations as you can for this. (e) Soils in tropical regions usually contain less humus than do soils in cooler climates. Explain.

3. Write a report on a community of your own choosing which points out the variety of strata in it. Include the components of each stratum and their respective niches.
4. Design and carry out an experiment to show the possible need for soil microflora and fauna in the growth of plants. You may be interested in limiting your investigation to non-parasitic nematodes or including all microorganisms.
5. Prepare a soil analysis report for pH (acidity), nitrogen, phosphorus, and potash for a number of locations on the school grounds.

RECOMMENDED READING

Applegate, Vernon C. and James W. Moffatt. 1955. *The Sea Lamprey.* Sci. Amer. 192(4): 36-41.

Avery, G. S. 1957. *The Dying Oaks.* Sci. Amer. 196(5): 112-122.

Bott, T. L. and T. D. Brock. 1969. *Bacterial Growth Rates Above 90° C in Yellowstone Hot Springs.* Science. 164 (3886): 1411-1412.

Farmer, D.S. 1950. *The Annual Stimulus for Migration.* Condor. 52:104-122.

Gilbert, Perry W. 1969. *The Shark: Barbarian and Benefactor.* BioScience. 18(10): 946-950.

Golley, F. B., J. T. McGinnis, R. G. Clements, H. I. Child and M. J. Duever. 1969. *The Structure of Tropical Forests in Panama and Colombia.* BioScience. 19(8): 693-697.

MacKinney, A. L. 1929. *Effects of Forest Litter on Soil Temperature and Soil Freezing in Autumn and Winter.* Ecology. 10:312-321.

Poulson, Thomas L. and W. B. While. 1969. *The Cave Environment.* Science. 165(3897): 971-981.

Read, Kenneth R. H. 1969. *Inshore Marine Animals in the Gulf of Maine.* Oceans. 1(3): 34-41.

Russell, C. P. 1932. *Seasonal Migration of Mule Deer.* Ecol. Monogr. 2:1-46.

Stephens, William M. 1969. *Symbiosis.* Oceans. 1(1): 70-75.

Went, F. W. 1955. *The Ecology of Desert Plants.* Sci. Amer. 192(4): 68-75.

Wilce, Robert. 1969. *A Look at New England Seaweeds.* Oceans. 1(3): 42-49.

7

"Let us hope that every effort is made to learn the structure, composition and dynamics of the original communities of North America before they finally succumb before the advance of civilization."
Shelford*

The Face of North America

Thus far, we have been looking at the details of communities – how they function and maintain themselves. Examples have perhaps helped in getting the over-all picture, but the ecological principles are only abstractions until they are used in binding together the diverse parts of a community. It is often difficult to apply concepts to real situations in a meaningful way. In the next few pages we will look at the faces of North America. The hope is that some loose strings will be tied together and that principles can be applied to the biological communities of which we are ultimately a very important part.

THE GEOGRAPHIC FACE OF NORTH AMERICA

The particular biology of any region is a result of many factors. The region's physical geography (or topography) is perhaps one of the most important of these factors. North America presents a vast array of contrasting geographic features and an array of biota to match. The physical features are also responsible, in part, for the climate diversity which may be found throughout the continent.

*Shelford, V. E. 1963. *The Ecology of North America.* The University of Illinois Press, Urbana.

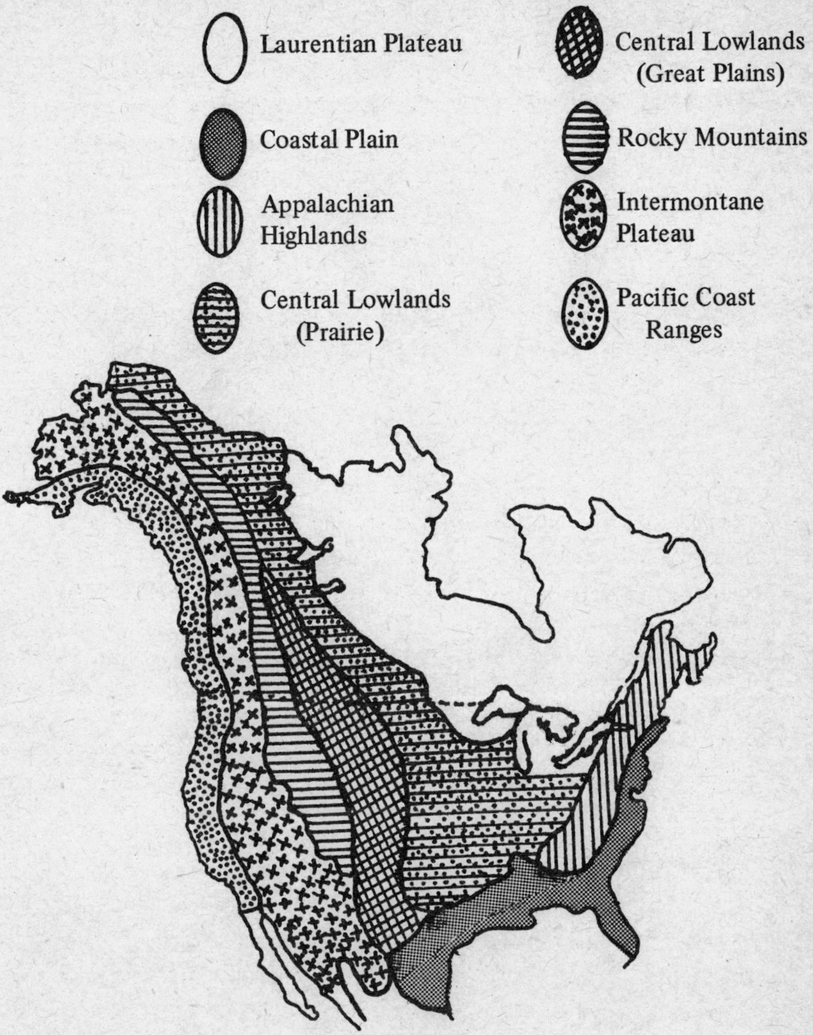

Fig. 7-1. The physical face of North America. The continent may generally be divided into several distinct regions based upon its physical features. (From Hylander, 1966.)

Looking at the overall picture of North America (Fig. 7-1), we see that it is a massive land area, slightly more narrow at its southern extremity. It is bounded on the east and west by two oceans, the Atlantic and Pacific, respectively. The northern portions of the continent stretch into the Arctic, while the southern portions are bathed by a subtropical sea, the Gulf of Mexico.

The eastern and western edges of the continent are sharply elevated into mountainous rims. Stretching between these two rims is a broad central lowland. The western continental rim is made up of two parallel mountain ranges, the Pacific Coast Ranges and the Rocky Mountains. This broad, rugged highland is over a thousand miles wide in some places, with towering, majestic peaks, some of which exceed 14,000 feet in elevation. The eastern rim, by contrast, is much less rugged, is much narrower, and not nearly as long. In addition, the western mountain rim drops more or less suddenly into the Pacific, offering little coastal plain. The eastern rim is set back from the coast and except for a relatively small area in the northern parts, offers a broad coastal plain between the mountains and the Atlantic Ocean. The Appalachian mountains, which comprise the eastern rim, are extremely old mountains, and in geologic times were some of the tallest. But time and weather have taken their toll in wearing away the once-lofty peaks. Today, they serve as a reminder of things to come for the much younger western mountains.

The two parallel ranges of the western rim are separated by a highland plateau. In Alaska, the northern extremity of both ranges, the plateau is narrow. It begins to broaden out in southern Washington, and constitutes the broad Intermontane Plateau between the Sierra Nevada Mountains to the west and the Rockies to the east. This highland plateau extends over much of Nevada, Utah, Arizona, and New Mexico.

The Central lowlands form a broad trough from north to south between the continental rims. It begins at the foot of the western mountains and slopes off toward the north, east and south, eventually meeting the Arctic Ocean, Appalachian Mountains and the coastal plain, respectively. The coastal plain separates the Atlantic Ocean from the Appalachian Highlands. It continues south, swinging westward in a broad arc south of the mountains into southern Texas as it meets the Gulf of Mexico.

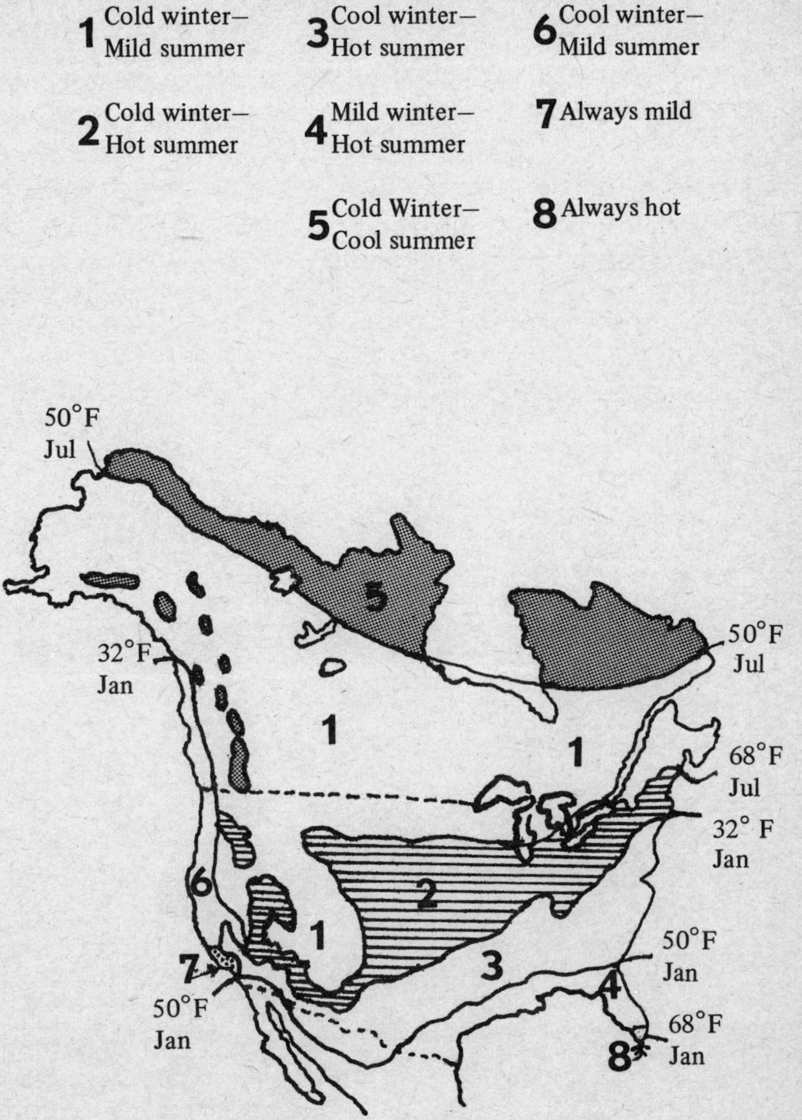

Fig. 7-2. The temperature belts of North America. Note the relationship between temperatures and various geographic features such as oceans and mountains. (From Hylander, 1966.)

Just as the physical face of North America is varied, so also is the climate. Figures 7-2 and 7-3 summarize the climatological face of our continent. The physical geography of the continent plays an extremely important part in determining the climate for any given region. For this reason, you should quickly notice the correlation between the temperature and precipitation belts and the topography. Latitude also plays an important role in climate. In the higher latitudes, the heating effects of the sun are decreased as the rays become less and less direct. (cf., Chapter 5).

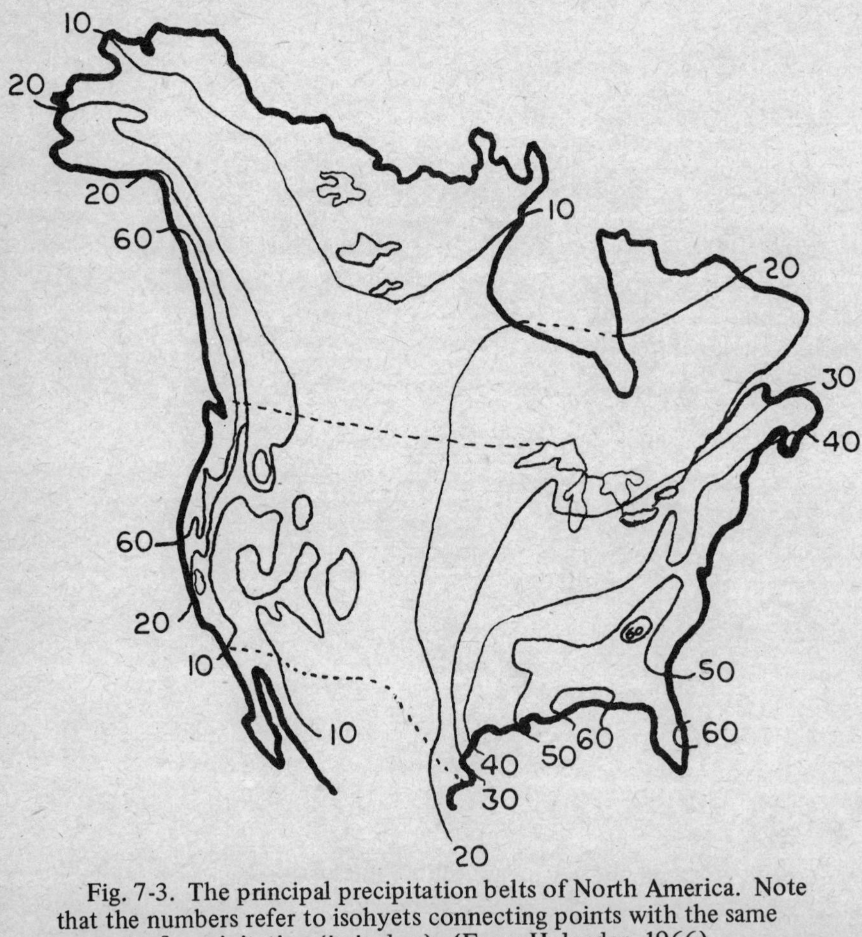

Fig. 7-3. The principal precipitation belts of North America. Note that the numbers refer to isohyets connecting points with the same amount of precipitation (in inches). (From Hylander, 1966).

If the average temperature for the warmest month of the year is below approximately 50°F., trees cannot survive. Consequently, the 50° temperature belt is of ecological significance and identifies the most northern belt in Figure 7-2. In effect, this also forms a boundary between areas which have a cold winter and cool summer and those with cold winters and mild summers.

Another important temperature line or *isotherm* is that which can be drawn for 68°F. during the warmest month. It is at this approximate temperature that conifer forests become limited. Thus, the boundary between regions of cold winter-mild summer and cold winter-hot summer is marked. The latter belt has its greatest expanse in the central United States, with a wedge extending eastward into New England. Another wedge wraps around the southern extent of the Rockies, reaching into the high desert of the Intermontane Plateau. The southern extent of this belt is the isotherm for 32°F. during the coldest month. The ecological importance of 32°F. is rather obvious.

South of this isotherm lies the cool winter-hot summer belt. Nearly all of this belt lies in the southeast to southcentral United States with a narrow finger extending into the broad California valley between the coastal hills and the high Sierra Nevada Mountains.

Following the southern Atlantic and Gulf coasts of the United States, the isotherm for 50°F. during the coldest month marks off the mild winter-hot summer belt. Except for the extreme tip of Florida, this belt occupies the remainder of the United States. The 60°F. isotherm for the coldest month marks the northern extent of the only tropical area of North America. The hot climate throughout the year provides an ideal environment for the lush vegetation and tropical animals which are limited elsewhere in North America.

The precipitation belts of North America have a more or less north-south orientation. This is a result of two interacting factors. First, the general weather pattern of the continent is west-to-east. Moisture is picked up from the Pacific Ocean and carried eastward. The continental land mass closest to the ocean receives the major portion of this moisture as expected. With less moisture available, less precipitation occurs as the air mass moves over the interior portions of the continent.

The second factor involved is the generally north-south arrangement of the topography. Air moving eastward is forced to rise over high mountain ranges. In doing so, it is cooled and the moisture which it carries is caused to condense. Consequently, the western slopes of the mountain ranges forming the western rim of North America receive most of the moisture in the air mass. When the air moves on to the eastern slopes of these mountains, it is much drier and warmer.

As the relatively dry air mass continues to move eastward, it picks up moisture from the land mass below. In addition, after passing onto the great Central Lowlands, the air mass is subjected to movements of moist air masses off of the Gulf of Mexico. This results in increased precipitation as the combined masses continue their easterly movement.

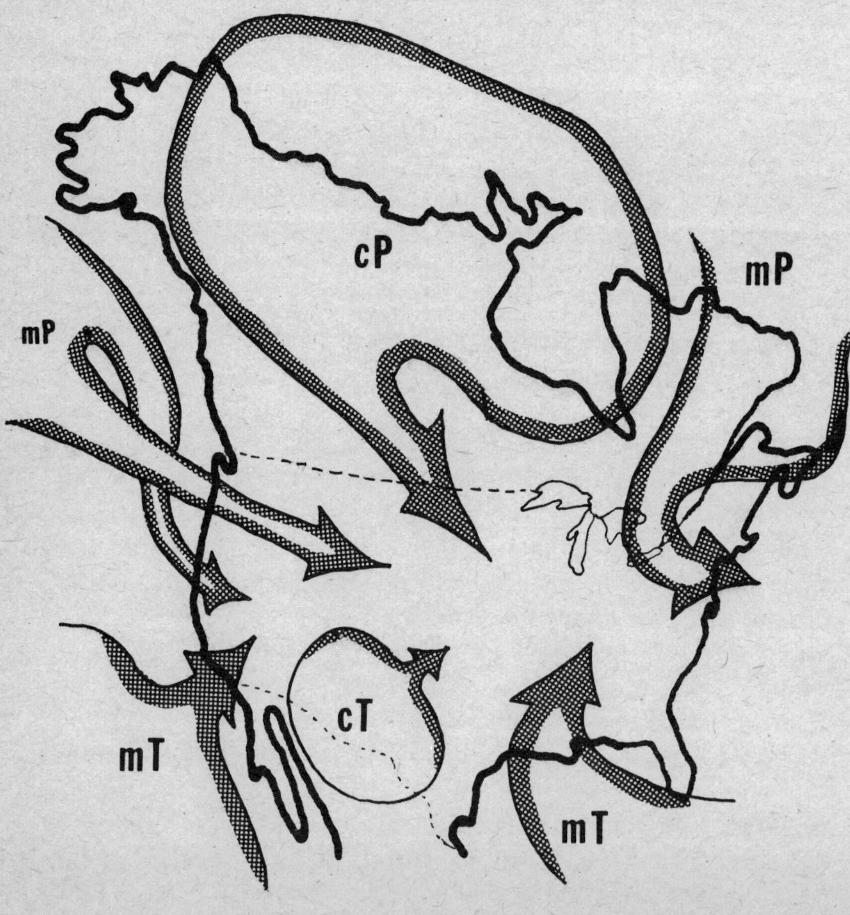

Fig. 7-4. Major air mass types which affect North America.

There are four principal types of air masses that exert their effects upon North America (Fig. 7-4). The first two of these, *maritime polar* (mP) and *maritime tropical* (mT), originate over various parts of the oceans; hence their names. The other two masses, *continental polar* (cP) and *continental tropical* (cT), have equally descriptive names denoting their origins. Which masses have the most effect at any given time is largely dependent upon the season. During the winter, the continental tropical air masses move southward as the sun's heating effects are greatly diminished in the northern hemisphere. Large polar snowfields form and spawn cold air masses which are able to descend upon the remainder of the continent.

In the summer, the heating effects of the sun are enhanced and reduction of the northern snowfields occurs. The tropical air masses are then able to exert a greater influence upon the climate of North America as they shift northward. These four masses are constantly interacting. They are ultimately responsible for the tremendous variety of weather which can occur throughout our continent.

Why does weather tend to move as it does? The answer to this question involves some elementary physics and a knowledge of the effects of the earth's rotation on its axis. The sun is again the prime mover – just as we saw the sun as the ultimate energy source for all biological activity in Chapter 3.

The global movement of air begins at the equator where solar radiation produces the greatest amount of heating. The heated air expands, becoming less dense, and is thus caused to rise. As it rises, it cools, but it cannot fall because more rising air from below continues to force it upward. The cooling air is forced to spill poleward. Being cooled, it also becomes more dense, and after leaving the equatorial belt, responds to the constant gravitational pull of the earth. It falls earthward, thus completing a giant convection belt of air movement (Fig. 7-5).

Meanwhile, cool air is also sinking at the poles and is being forced to flow toward the equator as more air sinks. As it flows along the surface of the earth, it is warmed and eventually forced to rise. If there were no surface features to the earth or no "tilt" to the axis, this global air movement would appear as in Figure 7-5A. However, such is not the case. The tilt of the earth's axis causes the most direct solar radiation to shift northward from the equator between March and September, and southward during the remaining months. This shift results in seasonal variations in the temperate regions. In addition, as the cool polar air moves toward the equator after its descent from the upper atmosphere, it is slowly warmed. Eventually, it becomes sufficiently warmed to begin ascending before it reaches the equator. Similarly, as the equator-warmed air travels poleward, it cools sufficiently to begin falling en masse at about 30° latitude. The large rotating air belt, or cell, of each hemisphere is thus broken into three smaller cells. The middle cell is driven by the other two much as a gear train is moved (Fig. 7-5B). With this arrangement, a significant transfer of tropical heat poleward is accomplished in a gigantic "bucket brigade."

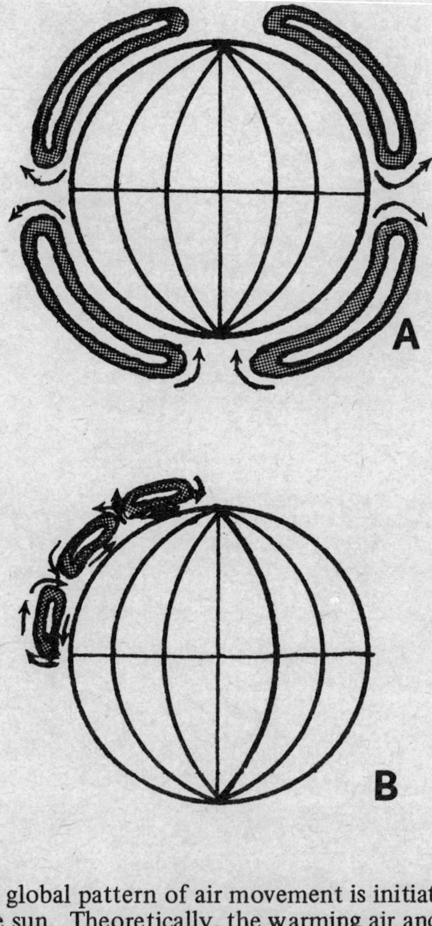

Fig. 7-5. The global pattern of air movement is initiated by the heating effects of the sun. Theoretically, the warming air and cooling air would cause a pattern as shown in A. Actually, the cycle for each hemisphere is broken into three cells as shown in B.

This relatively simple rotation of air cells becomes somewhat complicated by the rotation of the earth. If this were not so, the cells would produce a south-to-north movement of air in North America, instead of the more west-to-east movement which which we actually observe. In the northern hemisphere, any moving object that is not moving parallel to the direction of rotation of the earth is caused to veer to the right. The reverse is true for moving objects in the southern hemisphere.

This deflection of moving objects was first described in the 19th Century by a French mathematician, G. G. Coriolis. The so-called *Coriolis Effect* describes the deflection of moving air masses which result in the various trade winds (Fig. 7-6). Those winds with which North America is most concerned are the *prevailing westerlies* which produce the generally west-to-east movement of air masses across our continent.

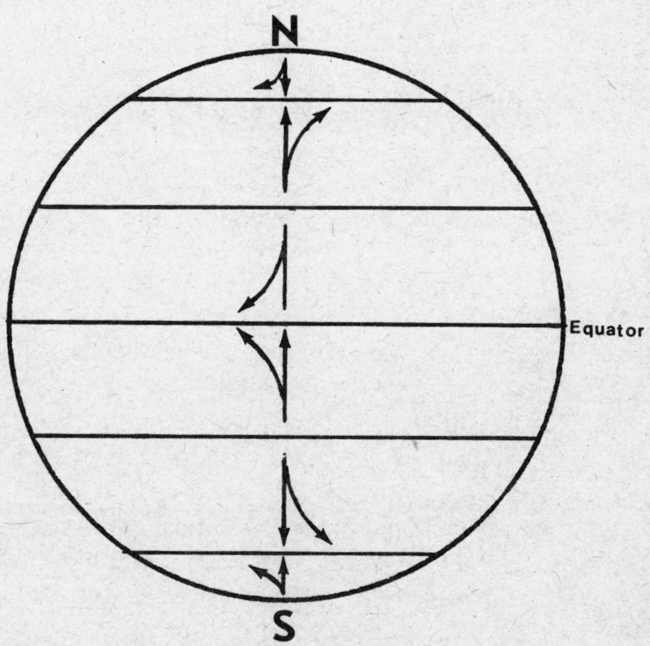

Fig. 7-6. The rotation of the earth produces an effect known as the Coriolis Effect. Air movement in the northern hemisphere is deflected to the right. A left-deflection occurs in the southern hemisphere. These winds are familiar as trade winds.

If you were to take a trip from the west coast of the United States through the middle of the country to the east coast, your journey would take you through nearly every principal geographic region of the continent. The combined effects of topography and climate would be sharply illustrated as you passed from one region to another. A summary of your trip might appear as in Figure 7-7.

Note that the Pacific coast mountain ranges rise almost directly out of the ocean with little coastal plain. Moist air moving onto the land from the Pacific is forced to rise over the high Sierra Nevada Mountains. Upon rising, it cools, condenses and is forced to lose its moisture as rain or snow. The moist conditions produced by the Pacific air as it meets the Sierras is an important factor in the perpetuation of the famous redwood forests of California. The eastern slopes of the mountains are denied such an abundance of moisture, and the high desert of the Intermontane Plateau stretches to the Rockies, giving up what little moisture it possesses to the dried air mass as it moves eastward.

The Rocky Mountains again force the air to rise as it continues eastward. The air is cooled a second time in the process. Again, moisture is lost on the western slopes and a new dry air mass moves out onto the Central Lowlands. Moist maritime tropical air from the Gulf of Mexico exerts a strong influence upon air masses as they leave the Rockies, and precipitation is more plentiful across the prairies than in the Intermontane Plateau region. The tall grass prairies of the eastern Great Plains result.

Crossing the Mississippi River, the annual precipitation becomes sufficient to support vast deciduous forests which blanket the eastern one-third of the Central Lowlands. Air moving across the Appalachian Highlands is greatly influenced by the Atlantic Ocean and its moist air masses. Consequently, the broad coastal plain of the east coast is well supplied with moisture to support an abundance of life forms.

An ecological cross-section of the United States only points out the so-called typical features that one might encounter in a coast-to-coast trip. Localized conditions do exist in a large number of places. A typical cross-section cannot possibly pass through all of these areas. For example, fairly localized conditions produce such outstanding exceptions as the Olympic Peninsula in Washington, the Black Hills in South Dakota, the more

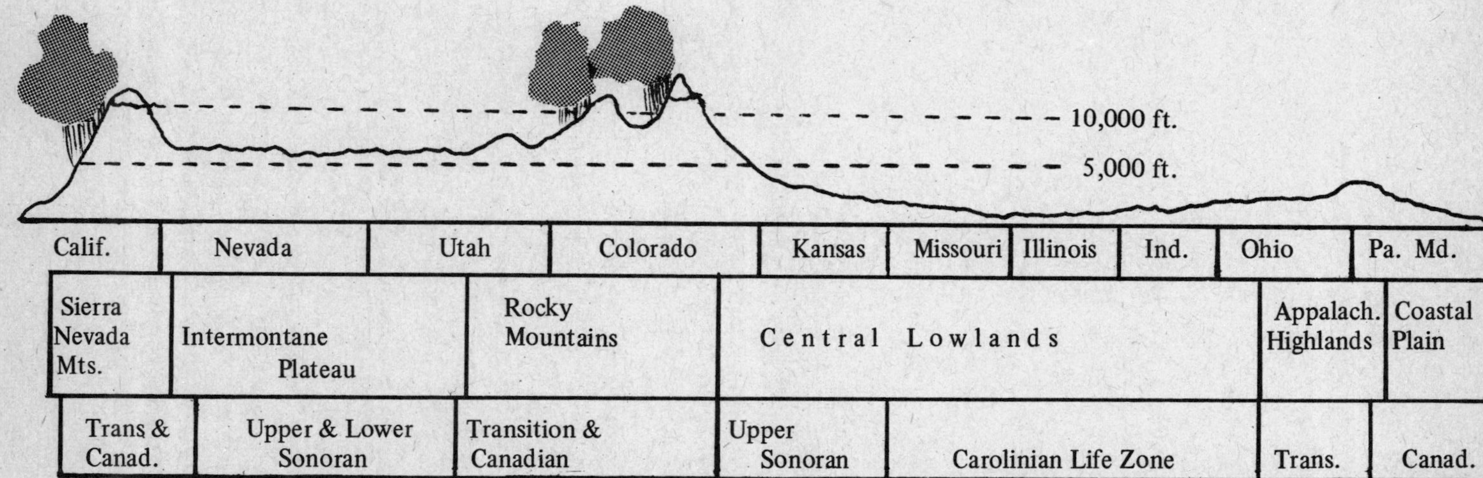

Fig. 7-7. An imaginary trip across the United States might take you through country that would be similar to this cross-section. (Adapted from Hylander, 1966.)

rugged parts of the Appalachian Highlands in New England, and the Ozark Plateau in southern Missouri and Arkansas.

Aside from being a constant source of moisture to adjacent continental areas, the oceans are important in another significant way. They exert a tremendous modifying effect on the seasonal weather that might ordinarily be expected at a given latitude. For example, the Great Plains at 45° N latitude normally experiences hot summers (where temperatures often exceed 100° F.) and cold winters (where temperatures often plunge below zero). But coastal areas in California at the same latitude or further north experience a great modifying effect by the ocean.

The reason for this is due principally to two very important properties of water. The first property is simply that water is a fluid. On land, solar radiation warms only the upper surface particles, not penetrating to great depths since the particles are not circulating and distributing the heat. No doubt you have experienced this fact on a hot day while walking on hot sand. You quickly discovered that you could avoid burning your feet by burying them only a few inches beneath the surface. But when water absorbs heat, it circulates throughout much of the body of water, heating much more than just the upper surface. Similarly, when it cools, the whole mass cools, not just the surface.

The second property of water which causes it to exert profound modifying effects is that it is capable of absorbing relatively large quantities of heat without significantly changing its temperature. During the summer months, the oceans are heated only a small amount compared to adjacent land masses. Consequently, the air over these cooler oceans remains relatively cool. Passing onto a land mass in California, this cool air tempers the heating effects of the sun that one might normally expect without the presence of the ocean.

In the winter, when the sun has a reduced heating effect, the oceans will slowly release the vast quantities of stored heat to the air above. This warmed air will again modify the cooler temperatures one might expect on the adjacent land areas. It should be apparent by now that the severe cold spells which the Central Lowlands experience are a result of continental polar air masses pushing southward. These air masses do not originate over water and are therefore much colder than those air masses which do.

The familiar on-shore breezes during the day and off-shore breezes at night are also an effect of unequal heating and cooling between water and land masses. During the day, the land surface becomes heated by the sun. Heat radiating from the land warms the air above it, causing it to rise. This is similar to the heating effects in equatorial areas. Cooler air from above the nearby water mass rushes in to take its place. The convection circulation thus creates an *on-shore breeze* (Fig. 7-8A). At night, the land cools quickly; but the water remains virtually unchanged in temperature. The water then serves as a heat source for the air above it, causing it to rise. The daytime convection circulation is then reversed, and an *off-shore breeze* occurs (Fig. 7-8B).

Unequal heating on various parts of the globe is actually the cause of all surface winds. The so-called sea breezes just discussed are only specific examples of a phenomenon that is actually the cause of winds everywhere. As air is heated, it rises. This creates an area of low pressure. When air cools, it falls in response to gravity. Where it descends, an area of high pressure is created. The air is forced away from the high pressure areas by more descending air and its natural direction will be toward the low pressure areas.

Ideally, then, winds would be created in a direction from high-to-low pressure centers. If it were not for the motion of the earth, this would be true. But the earth is moving, and the moving air is deflected in such a way that it tends to blow clockwise around a high pressure area and counterclockwise around a low pressure area (Fig. 7-9).

To summarize, the ecology of any given area is a product of its topography and its climate. The climate is directly related to the topographical features since it can be influenced by mountains and adjacent oceans or large lakes. The global air cells in conjunction with the rotation of the earth create the trade winds which influence the general global weather patterns. The fluid nature of water and its ability to absorb and release large quantities of heat without changing temperature contribute to the modifying effect of oceans on adjacent land masses.

Fig. 7-8. On-shore breezes (A) are created by convection currents set up by the sun's heating effects upon the earth. At night (B), the earth cools quickly and the air above the water becomes warmed by the residual heat in the water. A convection current is again set up, but in the opposite direction.

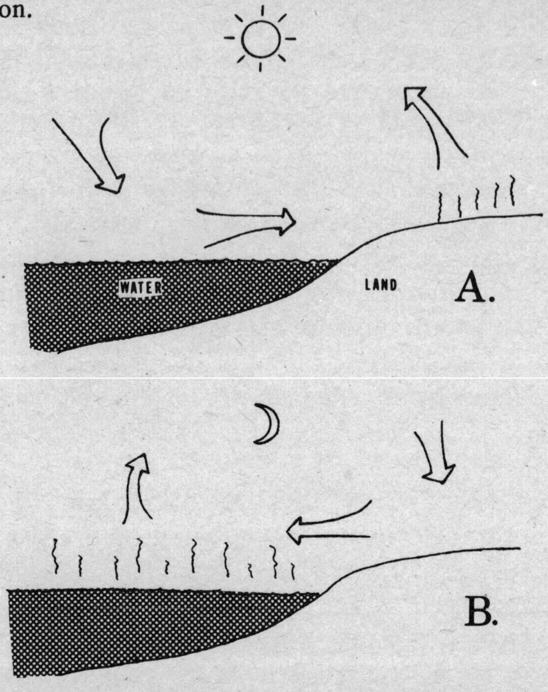

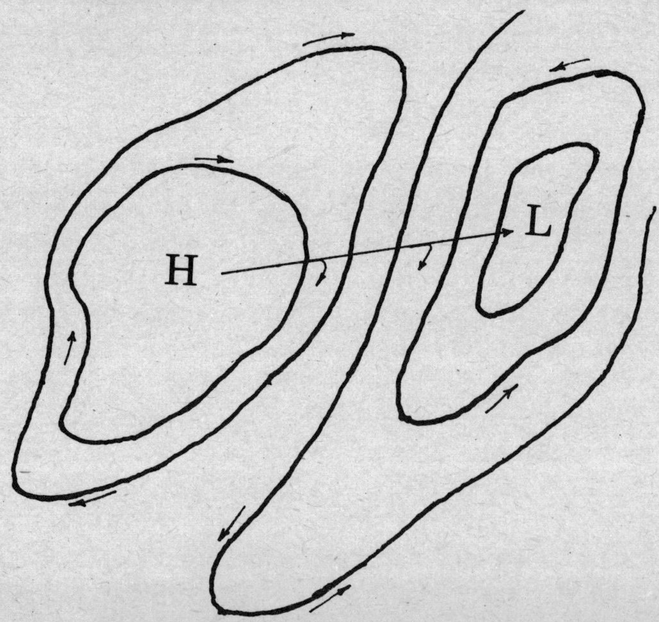

Fig. 7-9. Winds tend to blow from regions of high pressure to regions of low pressure (long arrow). But the rotation of the earth causes these winds to be deflected, creating clockwise air movement around a high, and counter-clockwise movement around a low.

We have spent a great deal of time discussing the general climatic features of North America and the causes behind them. To understand the biological distribution on the continent, we need to see it in the light of the climatic features which are present. Temperature and rainfall are the most noteworthy of the climatic features, and they consequently will be the most effective in determining the distribution of wildlife. The succession of temperature belts from north-to-south and the precipitation belts from east-to-west combine to produce eight major *life zones* on our continent.

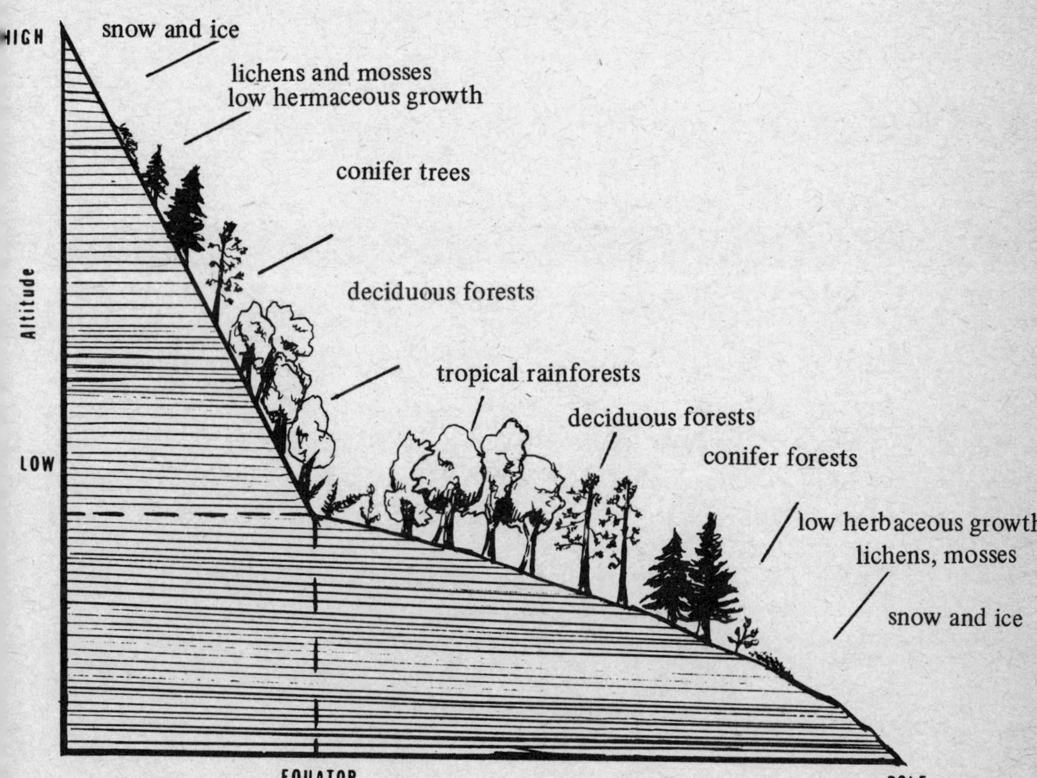

Fig. 7-10. Notice the striking similarities between vegetational zones correlated with altitude and those correlated with latitude.

It had always been common knowledge that there were definite biotic changes which occurred with changes in latitude. It was equally common that there were definite biotic changes associated with changes in elevation. But it was not until the early 19th Century that the world-traveling biologist Alexander von Humboldt pointed out the striking similarities between the biotic changes associated with latitude and those associated with elevation.

North America dramatically illustrates this concept in the western mountains. Figure 7-10 depicts the typical changes which can be observed with changing elevation and relates them to the latitudinal changes. Note that a change in elevation of 1000 feet is roughly equivalent to a change in latitude of approximately 500 miles at sea level.

The *Arctic-Alpine Life Zone* is represented in North America in two distinct areas. The first of these lies in the northernmost reaches of the continent. Bordered to the north by the Arctic Ocean, the high latitude allows only a very short growing season. The rigorous climate permits only the most hardy of species to survive. There is, consequently, not a great deal of diversity among the plants and animals represented. The plant forms are small, low-growing lichens, mosses and grasses. A few dwarf willow trees may be present giving some protection to the few birds and mammals that live in this frigid area. The second area which represents this life zone is found on the higher mountains of the western rim of the continent. Commonly referred to as the area above timberline, it is generally similar to the Arctic areas. Some of the higher altitudes may possess a perpetual cover of snow and ice.

Below the treeline on mountains as well as in northern Canada and Alaska lies the *Boreal Life Zone.* This area is characterized by spruce and fir trees in the northern (or higher) areas of the zone. To the south (or at lower elevations), ponderosa pine and Douglasfir are in greater predominance. The growing season lasts from three to five months. Animal life is more abundant also, with moose, wolverines, and large variety of birds, especially of the migratory type. The southern boundary of this life zone is the 68°F. isotherm for the warmest month. As noted above, warmer temperatures tend to limit the growth of conifer forests.

The *Transition Life Zone* forms a border between the Boreal forests and the milder winter areas to the south. Its greatest expanse is in the northern parts of the Central Lowlands. The growing season is six to seven months in length. Just as its name implies, it forms a transition belt between the conifer and deciduous forests. Much of the formerly forested areas of this zone are good grain-producing areas and have been cleared in the past 150 years to make way for agriculture on a broad scale. In the western regions, especially where the mountains produce a *rain shadow* resulting in semi-arid conditions, the Transition Life Zone is represented by pinon pine and juniper forests. Elsewhere, mixed conifer and deciduous forests occur if they have not already been cleared.

The fourth life zone is the *Upper Sonoran.* It includes the semi-arid areas of the Central Lowlands where annual precipitation is less than 20 inches. A narrow finger of this zone is represented in the Intermontane Plateau and around the base of the Pacific Coast Mountains. Vegetation is represented primarily by sagebrush and desert grasses. The more protected valleys may exhibit willows and aspens which provide shelter for the coyotes, deer and bird life of the Upper Sonoran region.

The *Lower Sonoran Life Zone* is even more arid. In this zone lie the great deserts of North America. Vegetation is represented by mesquite, a variety of cacti, and creosote bush. Animals, as well as plants, must be especially well adapted to the drying conditions of the desert. Lizards, snakes and small rodents are most abundant. Although the growing season is nearly year-long, the annual rainfall of only four inches or less comes usually in a short span of time, if at all.

Turning to more moist areas, the remaining three life zones are found in the eastern half of the United States. The *Carolinian Zone* is a region of cool winters and hot summers. It occupies the bulk of the Central Lowlands and averages some 40 inches of precipitation annually. The growing season may be as long as eight months. It is characterized by deciduous forests, the particular climax species being dependent upon the region. The typical dominant trees are magnolia, dogwood, tulip, birch and maple, but some higher areas such as the Ozark Plateau may characteristically exhibit oak and hickory.

South of the Carolinian Zone is the *Louisianian Life Zone.* Winters are mild and moisture is abundant, stimulating life forms such as amphibians, reptiles and birds to levels impossible in more rigorous areas. Much of the region is low-lying, near sea-level. Poor drainage results in the cypress swampland often stereotyped for this zone. Higher ground exhibits oak and pine forests.

The final life zone represented in North America is the *Tropical Life Zone.* Although not extensive on this continent, its biology is especially unique. The Everglades National Park is an area within this life zone that has been set aside to preserve the tropical features for the thousands of annual visitors. The growing season is not limited, being bounded on the north by the isotherm for 68° F. for the coldest month. Precipitation is abundant. The vegetation is typified by mangroves and palms; the fauna include exotic birds and crocodiles. With an abundance of niches, there is a corresponding abundance of life forms to fill them. With the exception of oceanic coral reefs which are not far off of the North American continent, the diversity of flora and fauna is the greatest.

The concept of life zones was the first good attempt to present a meaningful scheme that would lend understanding to the distribution of plants and animals. It was presented in 1892 by C. Hart Merriam. Although somewhat changed by some of the more modern ecologists, the basic structure of his ideas still stands. The life zones were based primarily upon temperature as the significant criterion. But since temperature is closely associated with many other ecological factors such as altitude, latitude, slope, exposure, prevailing winds, nearness to water and mountains, and type of soils, it is a fair environmental indicator.

However, the use of life zones to explain distribution, as do all schemes, has some important criticisms. Perhaps the most significant of these is that the average temperature calculations are made only for the normal growth and reproduction period. Often, we find that this is not the only important period. Not only are conditions during other parts of the year important, but daily and seasonal extremes can often be more important than growing season averages.

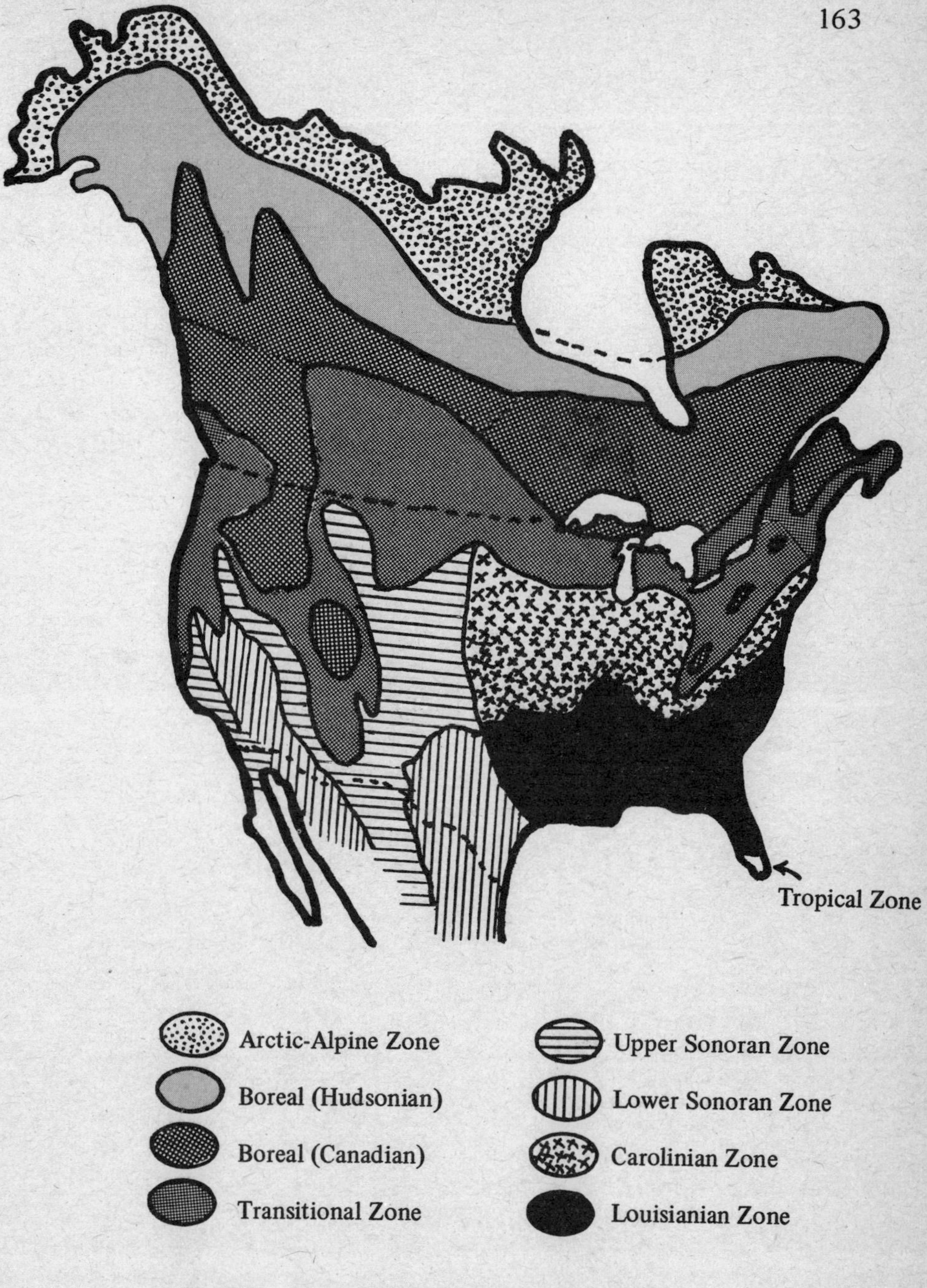

Fig. 7-11. Major Life Zones of North America.

Table 7-1. A Summary of the Life Zones of North America.

	Indicator Organisms		
Zone	Plants	Mammals	Birds
Arctic-Alpine	dwarf willow mosses lichens	arctic fox arctic hare lemming	snowy owl snow bunting
Boreal	Englemann spruce fir Douglasfir Ponderosa pine	wolverine woodland caribou marten lynx varying hare	great gray owl pine grosbeak spruce grosbeak Canada jay white-throated sparrow
Transition	birch maple Ponderosa pine	ground squirrel	sage grouse sharp-tailed grouse
Carolinian	beech maple magnolia tulip	fox squirrel prairie vole eastern wood rat	warblers lark sparrow cardinal
Upper Sonoran	sagebrush chapparal	grey fox ring-tailed cat grasshopper mouse	valley quail scrub jay titmouse
Louisianian	pine cypress	cotton rat rice rat	boat-tailed grackle chuck-wills-widow Swainson's warbler
Lower Sonoran	mesquite cactus creosote bush	kangaroo rat pocket mouse	desert quail cactus wren
Tropical	palm tree orchids mangroves strangler fig	black bear raccoon marsh rabbit spotted skunk	egret great white heron caracara

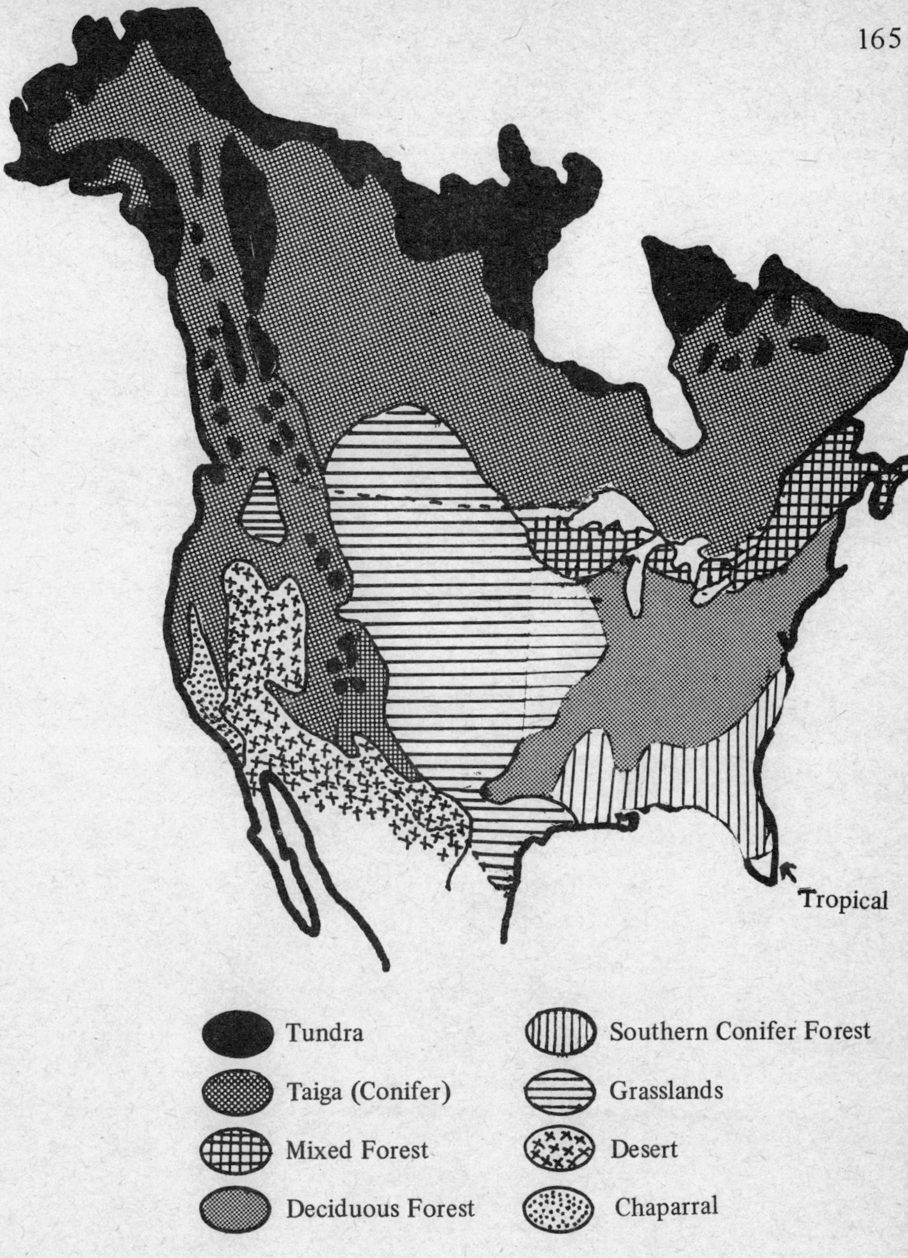

Fig. 7-12. Biomes of North America. (Adapted from Hylander, 1966.)

The alpine tundra is a treeless area with harsh winters and a short growing season. Plant variety is relatively low with short forms predominating. Even trees that grow to towering heights at lower elevations are dwarfed.

BIOMES OF NORTH AMERICA

Perhaps the most commonly used distribution scheme among ecologists today is the biome concept. It was first presented by F. E. Clements and Victor Shelford in 1939. The biome, in contrast to the strict temperature criterion of the life zone concept, is based upon a single plant type formation and its associated animals. Figure 7-12 outlines the biomes of North America. Each biome is a broad geographic area characterized by a single type of climate and the potential for a single type of climax community. The word potential, as used here, is important because the biome may actually contain several seral stages of a certain future climax community. The particular climax, remember, is subject to the climate for a given area. In addition, a particular biome may also contain non-climatic climaxes. Such communities are extremely localized and are exemplified by the deciduous woodlands which line the stream banks coursing through the grasslands. A single biome may also contain several closely related climax communities. Ordinarily, the biome is named for the dominant floral types, such as the Grassland Biome. Although this normally implies certain faunal associations, some ecologists prefer a more descriptive naming technique. Hence, the Grass-Bison-Pronghorn Biome may be used to describe the same region.

THE TUNDRA BIOME

The greater part of North America north of 57° N. latitude is referred to as the Tundra Biome. Even during the summer months, the sun's angle is not high, and the heating effects are low. Even so, the long days somewhat make up for the reduced intensity of the radiant energy reaching the area. The strong winter winds restrict plant life to low-growing forms – lichens, mosses, grasses, sedges and dwarf willows. Mean temperatures range from -20° to -26°F. during the coldest month (January) even at the southern boundary of the biome (see climatogram for Pt. Barrow, Alaska, Fig. 7-13). Precipitation, falling mostly as snow, is relatively low, ranging from 12 to 20 inches annually, decreasing northward.

The Arctic tundra is generally a low-lying, flat or gently rolling plain. During the short summer months, only the uppermost layers of the ground are warmed sufficiently enough to melt the frozen soil water. Drainage is poor because the melting snow cannot sink below the *permafrost* (that part of the ground which is permanently frozen). Numerous lakes, ponds and marshes fill the lower valleys and depressions. Thick mats of peatlike material full of roots and stems of tundra vegetation produce bog-like conditions.

When winter comes, the summer's end is quick. Ponds freeze over, often to the bottom. The sun's angle decreases steadily until it fails to rise above the horizon. At the Arctic Circle, this occurs only one day out of the year, but if you were to progress northward, the period during which the sun would fail to rise would steadily increase to six months at the North Pole. The lack of sun as well as the relatively high latitude results in extremely cold temperatures. Animal life is hampered by a scarcity of food and the rigorous physical conditions. Most are forced to migrate to more favorable areas. The caribou move into the forested regions to the south. Large numbers of birds, especially waterfowl move much farther. Other animals move only as far as the seashores where they are able to take up a commensal relationship with the polar bear. Few species are able to survive such inhospitable conditions, and the overall diversity of species is consequently very low.

The Alpine tundra is that barren highland above treeline. Treeline varies in altitude from south to north, that closer to the equator being considerably higher than in Alaska. The Alpine tundra occurring in Alaska occurs as low as 3,500 feet. Above treeline on Mt. McKinley, where precipitation is only approximately 25 inches annually, the soil is permanently frozen on the north-facing slopes. In Colorado, Alpine tundra occurs above approximately 11,000 feet, and few areas retain snow or frozen ground permanently. Usually, only drifts of snow remain throughout the summer on north-facing slopes or in the shadow of overhanging rock faces. Alpine tundra vegetation is not unlike that of the Arctic. The topography, of course differs markedly. Animal forms also are in contrast, especially with respect to the larger mammals such as polar bears, musk ox, and wolves which do not ordinarily occur in the high mountain regions. Caribou occur in the high tundra of Alaska, during the harsh winters. The Alpine tundra is populated by mountain sheep, mountain goats, marmots, ground squirrels and mice in

The taiga biome is a vast forest of Douglasfir, blue spruce, Englemann spruce, and other conifers.

various parts and in varying frequencies. The ptarmigan is also a year-round resident of the high tundra.

THE TAIGA BIOME

South of the tundra (or at lower elevations), the climatic conditions are sufficiently mild to support the growth of trees. The structure of the boreal coniferous forest varies widely throughout this area, often referred to as the Taiga Biome. Solar radiation is much greater than that received in the tundra, even though the summer days are shorter. Permafrost is virtually nonexistent. Although the temperatures do not reach the extreme cold levels which are common in the tundra, snowfall is considerably greater. Debris left from the retreat of glaciers in the past twenty thousand years provides natural dams which catch the run-off from the snow. Every year, thousands of sportsmen come to the coniferous forests of the Taiga to hunt, fish and canoe the vast wilderness of northern Canada and its myriad of lakes and rivers.

The white spruce-balsam fir forest is the principal climax community of this biome. It extends from Newfoundland to the foothills of the Rocky Mountains. There are also numerous subclimax areas and/or seral stages which exist locally throughout the biome, some extending southward in the lower mountains, into the montane forests of the Rockies and Sierra Nevadas. Precipitation ranges from 10-20 inches annually in the eastern half of the biome to as much as 50 inches in the western regions. Mean temperatures may fall as low as -4° F. east of Lake Superior and -20° F. to the west. Maximum summertime temperatures may attain as much as 100° F. or occasionally higher. The overall climate, however, may be described as cool to cold and moist (see climatogram for Moose Factory, Canada, Fig. 7-14).

Photosynthesis is limited primarily to the upper reaches of the trees. Understory vegetation is sparse. Occasional fires or windstorms may destroy a forested area allowing a dense growth of typical disturbed-area species such as birch, aspen, and a variety of ferns and herbs. These are soon replaced in typical successional pattern by the conifers of the particular climax species. Insects are abundant in the Taiga. Insectivorous birds are consequently also available in abundance during the summer months, migrating southward with the approaching winter. Many seed-eaters, however, remain throughout the winter, although even some of these species may migrate short distances.

DECIDUOUS FOREST BIOME

As we continue to move southward, the conifer trees of the Taiga Biome begin to give way to deciduous trees. In the northern parts of the eastern half of the United States, the increased rainfall, warmer temperatures and different soils provide a combination of factors which allow climax communities of beech and maple to develop. The long days of the summer coupled with the relatively direct rays of the sun combine to provide enormous amounts of solar radiation. Although the annual radiation received by this area is much lower than tropical areas, the summertime radiation is much greater. Weather is extremely variable. Permafrost is virtually non-existent. Winter months may be very cold but often, the ground is clear of snow. Even mild temperatures of 60-70°F. may occur during the coldest months in some of the northern areas of this biome.

Animals may remain active through much of the winter. Photosynthetic activity is nearly absent since leaves are lost from the vegetation. But plants are still providing the food for the consumer community. A variety of berries, seeds and saplings all provide nourishment for the herbivores. Some must continually forage for such foods, but others have spent a great deal of time stocking a storehouse to provide food when it becomes scarce. Insects may hibernate under the bark of trees or in the insulating mat of decaying vegetation on the ground. These insects, in turn, provide food for the few insectivorous birds which overwinter here and can search out their food.

Several animal adaptations to winter have been discussed previously. Suffice it here to mention that such adaptations may take the form of migration, hibernation, dormancy and alteration of appearance. Plants, while they cannot migrate with seasonal changes, usually become dormant or actually die each season, producing seed for the following season. These latter plants are called *annuals,* and while they may be the first plants to establish themselves in secondary succession, they are

The grassland biome is generally treeless except where a river or stream courses its way to the sea.

usually a minority group in most communities. *Perennials* are herbaceous plants that remain alive in the soil through the winter and send up new shoots during the growing season. Survival of shrubs and trees is insured since they remain alive for a number of years and are protected by their woody stems. Should the growing season be shortened by weather or the seed crop of a given year be destroyed by insects, disease, fire or some other catastrophe, the species is protected against extinction.

Summertime in the deciduous forest reveals the presence of a great many niches. A canopy layer (c.f. stratification, Chapter 6) catches the full strength of the solar radiation. Although this layer is the primary photosynthetic area, much light filters between and through the leaves for the understory layers and even for ground-level shrubs, herbs, ferns and mosses. Each layer of a forest community possesses a distinctive group of animals. The treetops offer a set of environmental conditions that is uniquely different from lower levels. The cool, moist forest floor is ideal for a variety of woodland snakes, frogs, salamanders, birds, and a hoard of insects. Skunks, raccoons, mice, and deer make up the bulk of the mammal population although occasional bears, wolves or mountain lions are sighted.

The particular character of a local deciduous forest is widely varied depending upon its location. Precipitation may vary from as little as 25 inches annually in the western portions of the biome to as much as 80 inches in the southern Appalachian highlands. Temperature may be relatively mild in some areas, but display a wide seasonal range in others. The result is a biome characterized by a number of different communities, each with its particular species of plants and animals. While some species are widely distributed throughout a variety of different communities in the biome, many with smaller tolerance ranges are more locally distributed.

THE GRASSLAND BIOME

The Great Plains of the North American continent make up the Grassland Biome. While the character of this biome is also varied with respect to particular community dominants throughout its extent, it nevertheless possesses several unifying features. Precipitation is generally less than 20 inches annually. Much of this falls as snow although there are often long periods of bare ground in the winter.

The eastern portions of the grasslands may receive as much as 35-40 inches of precipitation annually. In spite of the adequate moisture for the development of tree-dominated communities in the eastern portions of the prairie, other physical factors limit their growth. Frequent droughts and wind-swept fires were important in the preservation of the prairie before the advent of the white man in the area. Excessive transpiration caused by constant winds, is always a barrier to the spread of trees. Grasses are well-adapted to such conditions, and the growth of tree-dominated communities is limited to fairly localized areas such as sheltered ravines. Until the coming of the pioneer into the Great Plains, grasses in high precipitation portions of the prairie were taller than a man's head. It was often recorded that people became lost after having wandered away from their camp because they could not see out to get their bearings straight. Civilization has since managed to methodically remove the original prairie and replace it with cropland where vast fields of wheat, corn, barley, oats and other grains are now raised.

Grasses are members of a large group of herbaceous perennials which dominate this biome. They possess several rather unique characteristics which make them especially well-suited for life in the rather rigorous environment of the open prairie. One of the most unique of these adaptations is their manner of growth. Each leaf blade grows continuously from its base – a distinct advantage to a plant that is continually being grazed upon by the herbivorous elements of the community. Most plants grow from the tips of their stems and can be easily ruined by the removal of this portion of the plant. As long as enough of the leaf blade remains to carry on photosynthetic activity, a constant food supply is maintained for the plant.

Blades of grass also grow in such a manner as to possess a large surface area for each plant without overshadowing neighboring leaves. They are elongated and grow more or less vertically which enables them to make maximum use of the sunlight. At the same time, the vertical arrangement protects the plant from the extremely high intensity of the solar radiation to which it is constantly exposed.

The roots, too, offer a special adaptive feature. Many roots of grasses in areas of low precipitation not only have a set of

Fig. 7-13. Tundra

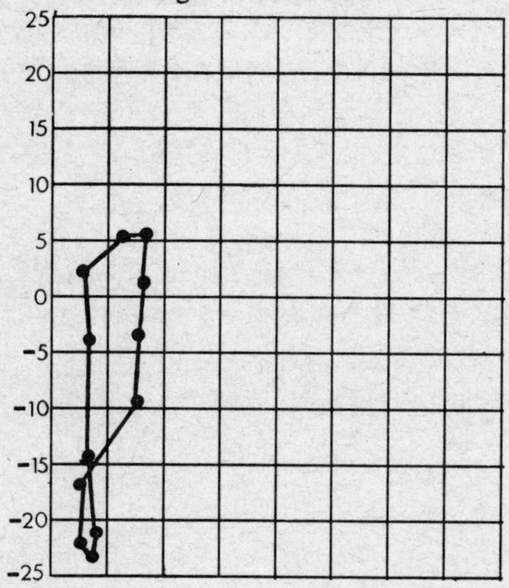

Fig. 7-14. Taiga.

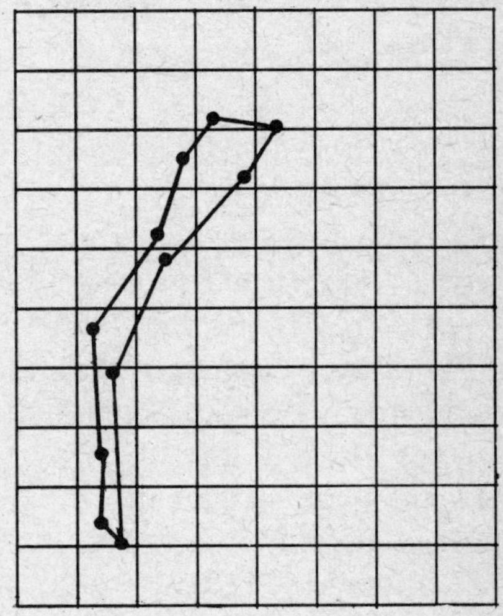

Fig. 7-15. Deciduous Forest

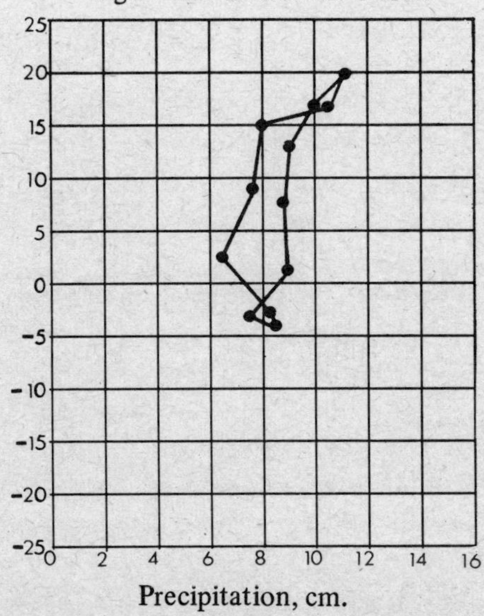

Fig. 7-16. Grassland

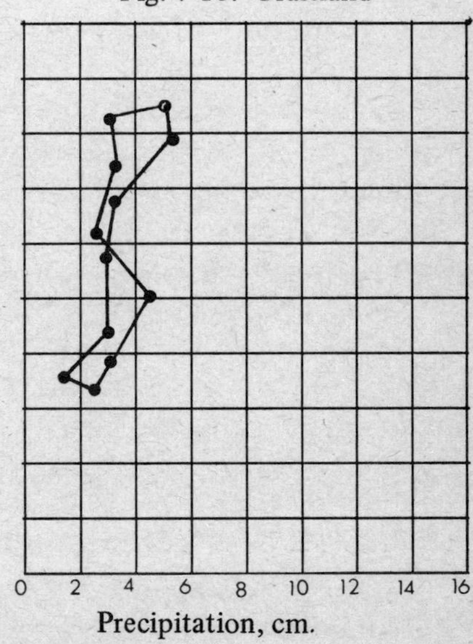

deep roots to search out the deeper water supplies unavailable at the surface, but they possess a fine network of surface roots as well which spread out in a dense mat just beneath the surface of the ground. This enables the plant to quickly absorb a maximum amount of the moisture from a small shower. When surface water supplies have been exhausted by transpiration and evaporation, the deep roots enable the plant to continue to find moisture.

Grasses are also well-known for other water-preserving adaptations. Small hairs on leaf surfaces are common features. These aid in retarding water loss by transpiration. A layer of higher humidity is created next to the leaf surface and evaporation is reduced. Some leaves curl when water is scarce. Special cells ordinarily kept tight by water prevent the leaf from curling when water is plentiful. When water becomes more scarce, then these cells lose water and become more flaccid, allowing the leaf to curl.

Herbivores might well be expected to be the dominant consumer on the prairie. The native populations of bison and pronghorn have been essentially replaced by cattle although attempts are being made to encourage their return. Less conspicuous are a great many burrowing rodents such as prairie dogs, ground squirrels and numerous species of mice. Jackrabbits and insects, especially grasshoppers are also common as well as a variety of game birds such as pheasant, bob-white quail and prairie chicken.

Predators common to the grasslands include coyotes, foxes, snakes, badgers, hawks, owls, numerous insectivorous birds, and at one time, wolves. This area is often the winter home of migrant seed-eating birds from areas farther north.

THE DESERT BIOME

There is no real dividing line between the short-grass prairies of the Great Plains and the more xeric areas which we call the desert. Some ecologists prefer to point out a semi-arid grassland area characterized by some grasses, cacti, and yucca. For the sake of simplicity, we will recognize this area as a broad ecotone region, possessing characteristics of grasslands as well as the more arid regions generally defined as desert.

The desert is often stereotyped as being hot, dry, and almost inactive biologically. However, there are few areas where

Where precipitation is less than 10 inches per year, the desert environment establishes a rigorous set of requirements for survival for living forms.

this is actually the case. North America possesses several types of deserts, the differences being a result of altitude, latitude, and location with respect to the mountains. In general, one could summarize the abiotic environment by saying that rainfall is scarce and evaporation is rapid. Soils are rather sandy and often possess a hard-pan fairly close to the surface. When precipitation does occur, it is likely to be brief, although possibly heavy. The hard-pan prevents the moisture from soaking into the ground, and consequently, run-off is rapid. The small amounts of vegetation cannot significantly retard the run-off and must be adapted to retaining as much of this water as possible.

Being situated in the western and southwestern portions of the continent, the deserts are constantly subjected to rather dry air masses. The reasons for this have already been explained earlier in the chapter. Because the air is so dry, heat loss from the ground at night is much more rapid than in more humid areas. The nights, therefore, are quite cool, often dropping to approximately 50°F. after a daytime high of over 100°. Temperature variations from day to night and season to season are extremely wide.

Vegetation in the desert possesses many unique variations to enable it to withstand the rigors of the environment. Transpirational surfaces are often reduced or absent. The spines of the cactus represent one adaptation to preserve moisture. These modified leaves also protect the plant from being grazed, thus serving a dual role. Since the rainy season is rather short (and often unreliable) plants must be able to respond to moisture quickly and complete their life cycles during this period. Plants are also usually widely spaced to prevent excessive competition for soil resources, especially moisture. Many plants secrete chemical inhibitors into the soil that prevent the germination and/or growth of other plants near them.

As might be expected, most plants are of the desert perennials. Besides the various types of cacti, these include mesquite and creosote bush. Xeric grasses also are present in the desert, expecially in the alkaline depressions which are readily indentified by caked salts on the surface. Animal activity in the hot deserts is generally confined to the twilight hours of morning and evening. Daytime termperatures are too high to allow activity to be carried on at a very fast pace. Food chains are short. Herbivores include peccaries, armadillos, skunks, a few insects, rodents such as the kangaroo and wood rats, various mice, rabbits and hares. Hawks, owls, coyotes, foxes, snakes, lizards and a few insectivorous birds represent the bulk of the predators.

The five biomes considered thus far occupy the greater part of the continent. Each biome occupies a rather broad area and while generalizations can be made as to its characteristics, one must recognize that variations do occur. These variations are the result of many factors including climatic, topographic, and influences of man. Nevertheless, these five areas can be rather easily identified.

There are, however, many localized situations that deserve brief mention before we leave the subject. These situations are not extensive in North America, but due to local conditions, several other biome types may be recognized. The tropical rainforest situation, for example, is closely approached in the Everglade National Park area of southern Florida. It has already been characterized as one of the life zones earlier in the chapter. Another unique rainforest deserves mention. Although not a tropical type, the Northern Pacific Forest such as that found on the Olympic Peninsula on the Washington coast exemplifies the mid-latitude rainforest situation. The climax vegetion here is the cedar and western hemlock. The Pacific Ocean provides an abundance of moisture (over 100 inches annually) and ensures a mild climate throughout the year despite the relatively high latitude.

The coastal plain and piedmont regions of southeastern United States are also somewhat unique. Some ecologists consider this to be merely a southern extension of the forest biome which covers much of the eastern part of the country. But the dominants are conifers such as loblolly, longleaf, and slash pine. The maritime forest communities along the Gulf Coast are dominated by live oak and cypress with long festoons of Spanish "moss" which is often stereotyped for much of this area.

The open, dry woodland of southern California extending into central and southern Mexico is also sometimes divided off as a separate biome. However, this *chaparral* area is not unlike the broad ecotone area separating the grasslands from the coniferous forests of the Rocky Mountains.

This is the Face of North America. It is as much varied throughout its length and breadth as anywhere in the world. But the variations can usually be traced to fundamental interactions of topography, weather movements, latitude and altitude as well as the frequent intervention of man. Much of this area is threatened today by the direct or indirect fumblings of man. It is of vital importance for us to become familiar with the complex interactions that go together to build a community in order that we don't in turn become the victims of our own fumblings. Pollution is not the problem that must be overcome. It is an attitude toward the environment that must be cultivated. With a healthy attitude and a proper respect for the forces of nature, our job becomes easier and the problems created by man will take care of themselves with little help.

QUESTIONS FOR DISCUSSION

1. Describe the changes that take place in an air mass as it moves off of the Pacific Ocean onto the North American Continent.
2. List the four principal air mass types and indicate their respective origins.
3. Describe why different air masses affect the central United States at different times of the year.
4. Describe the global movement of air cells in terms of the causes for such movement.
5. What is the Coriolis Effect? How does it affect the global air cells?
6. What are trade winds? How do they influence weather patterns in North America?
7. Describe the causes of on-shore and off-shore breezes.
8. In what important ways is water an influencing factor in climate?
9. In middle latitude biomes, many mammals hibernate in burrows; in the tundra, very few do. Explain.
10. Why is estivation unlikely in a tropical rainforest?
11. Describe some biome ecotones in North America.

12. By use of examples, point out why factors other than precipitation alone determine the prevailing climatic situation in any given place.
13. Explain why many species of migratory birds in Europe and North America are being found in higher latitudes than at the beginning of the century.
14. What is a biome? What is a life zone? How are these terms related?
15. Discuss briefly the differences between tundra, deciduous forest, grasslands, and tropical rainforest in terms of available niches. What is the relationship between the available niches and the diversity of species found in these areas?
16. What limiting factors exist in the Great Plains? in the true prairie?
17. Describe Alexander von Humboldt's latitude-altitude hypothesis. Explain why this is a generally valid hypothesis.

INVESTIGATIONS AND ACTIVITIES

1. Some geologists support the theory of *continental drift,* which holds that the present arrangement of continents is different from what it was in the distant geological past. Investigate this theory (several articles have appeared in *Scientific American* in the past six or seven years) and report on it as it applies to explaining the geographical distribution of plants and animals.
2. Choose some small taxonomic groups (at least three) that are present in your own state. Investigate their distribution in the world as a whole and construct maps to show this information. Try to explain the distribution in a brief report.
3. Make a climatogram of the region in which three animals or plants are normally found, and shade in the area lightly. For each organism, plot the points for a climatogram of a contrasting region *on* the first climatogram. Explain, in terms of your results, why the organisms live where they do and not (if, in fact they do not) live in the second region. If they do live in the second region, explain this.

4. Prepare a table of the dominant animals and dominant plants in each of the five biomes discussed in the text as well as two or three areas of special interest which are represented on the continent such as the Pacific Coast rainforest, tropical rainforest of Florida, and the chapparal.

RECOMMENDED READING

Bates, Marston. 1963. *Animal Worlds.* Random House, New York.

——— 1960. *The Forest and the Sea.* Random House, New York.

Beals, Edward W. 1969. *Vegetational Change Along Altitudinal Gradients.* Science. 165(3897): 981-985.

Benton, Allen H. 1966. *Field Biology and Ecology* McGraw-Hill, New York.

Elton, C. S. 1958. *The Ecology of Invasions by Animals and Plants.* John Wiley & Sons, New York.

Haag, W. G. 1962. *The Bering Strait Land Bridge.* Sci. Amer. 206(1): 112-120.

Hartesveldt, R. J. 1964. *The Ecology of the Giant Sequoias.* Natur. Hist. 73(10): 12-19.

Hylander, Clarence J. 1966. *Wildlife Communities from the Tundra to the Tropics in North America.* Houghton Mifflin Co., Boston.

Kilburn, P.D. 1965. *Floras of the Tundra.* Natur. Hist. 74(7): 52-59.

Leopold, A. S. and Editors of Life. 1961. *The Desert.* Time, Inc. Book Division, New York.

Pimentel, Richard A. 1963. *Natural History.* Reinhold Publishing Corp., New York.

Shelford, V.E. 1963. *The Ecology of North America.* Univ. of Illinois Press, Urbana.

GLOSSARY

Abiotic – the non-living aspects of an environment.

Absorption–the passage of one substance into another.

Adaptation–the adjustment of an organism to meet the demands of the environment.

Adsorption–collection of one substance upon the surface of another.

Air Cell–circulation pattern developed as air rises, cools, and descends to the earth.

Allen's Rule–the length of the extremities of warm-blooded animals is reduced proportionately as one travels poleward from the equator.

Annual–yearly; plants that germinate from seed, grow, flower and die within one growing season.

Arctic-Alpine Life Zone–life zone of the Arctic Ocean coastal plain; the mountainous areas above timberline.

Arhythmic–without rhythm; phenomena that occur with no particular cyclic character.

Autotroph–organisms that are capable of manufacturing their own food requirements from simpler raw materials, principally through photosynthesis; nearly all are green plants.

Bacteria–simple unicellular organisms found in air, soil, water; they may be free-living or parasitic.

Bergmann's Rule–the size of warm-blooded animals increases proportionately as one travels poleward from the equator thus reducing the ratio of body surface to volume.

Biogeochemical cycle–cyclic series of transformations of a chemical element through the organisms in a biotic community and their physical environment.

Biology–the study of life and its processes.

Biomass–the living weight of an organism.

Biome–large geographic area encompassing a major climax community type.

Biosphere–the layer of the earth and its atmosphere that contains all of the earth's life.

Biotic–pertaining to a living organism.

Biotic succession—natural replacement of one community by another.

Bog—community in which a terrestrial mat of vegetation has grown out over the surface of a pond, becoming thicker with age, but still often overlying the pond below.

Boreal Life Zone—characterized by coniferous forests, cool summers and cold winters.

Botany—the scientific study of plants.

Canopy layer—upper level of a forest community that forms a shading layer for lower community levels of the forest.

Capillarity—the elevation or depression of a liquid in a tube due to the attraction or repulsion of the liquid for the molecules of the tube.

Carbon Cycle—passage of carbon from atmospheric carbon dioxide into living tissue by photosynthesis and then through successive organisms until it is again released as carbon dioxide.

Carrying capacity—the maximum size of a population that an area can support indefinitely without undergoing deterioration.

Chaparral Biome—area of dense thickets of stiff or tough-leafed shrubs; characteristic of areas with Mediterranean climate such as southern California.

Chill factor—a value combining temperature and wind measurements that indicates what the temperature would feel like with no wind blowing.

Chloroplast—cellular body containing chlorophyll, a green pigment used by plants in the process of photosynthesis.

Circadian rhythm - daily rhythm.

Clear-cutting—removing all of the trees in a forest area.

Climate—generalization of weather conditions of a region throughout the year, averaged over a period of several years.

Climatic succession—replacement of one biotic community by another as a result of alterations in the climate of the region.

Climatogram—a graphic illustration of the climate of a given area in which monthly temperatures are plotted against monthly precipitation.

Climax community—the final stage of ecologic succession in an area.

Commensalism–symbiotic relationship between two organisms in which one species benefits and the other is neither benefited nor harmed.

Community–all of the organisms of all species living in a particular area or region.

Community dominant–the species of organism that is indicative of the community as a whole; species that is in greatest abundance.

Competition–a struggle between individuals of the same or different species for food, shelter, mates, space or any other limited resource.

Coniferous–belonging to a plant group, mostly evergreen, that include the pines, firs, spruces and other cone-bearing plants.

Conservation–wise use of natural resources.

Consumer–an organism that relies upon other organisms for its food materials, as opposed to a green plant.

Coriolis Effect–deflection of moving objects to the right in the Northern Hemisphere and to the left in the Southern Hemisphere brought about by the rotation of the earth.

Decay–the chemical breakdown of organisms and their products into simpler materials.

Deciduous–a group of perennial plants that lose their leaves at the end of each growing season.

Decomposer–an organism that feeds upon and breaks down dead organic matter.

Density–size of a population within a particular unit of space.

Desert – an arid region receiving less than ten inches of precipitation annually, characterized by high diurnal temperatures, and relatively low diversity of flora and fauna.

Diffusion–movement of a substance from a region of high concentration to one of low concentration.

Diurnal–daily.

Dominant–species of plant or animal that is particularly abundant or controls a major portion of the energy flow in a community.

Dormant–a state of inactivity; usually refers to extended winter sleep periods of certain organisms such as the bear.

Dynamic equilibrium—state of ecologic balance within an ecosystem in which imbalance introduced by activities affecting the community is offset by a readjustment of other activities within the community.

Ecologic equivalent—an organism that fills approximately the same niche in one community that another organism fills in another community.

Ecology—the study of the environment, its flora, fauna, and abiotic substances, and their relationships to each other.

Ecosystem—a biotic community and its physical environment.

Ecotone—transition area between two communities.

Ecotype – a locally-adapted population that is a result of natural selection.

Edge effect—blending of two communities to produce an ecotone.

Emigration—departure from one place to take up residence in another.

Energy—ability to do work.

Entomology—the study of insects.

Environment—physical and biologic characteristics of a certain area.

Epiphyte – non parasitic plant that grows upon other plants for physical support, deriving its required moisture chiefly from the air; includes such plants as arboreal orchids, bromeliads, and many mosses and ferns.

Estivation—period of minimal activity during the warmer and often drier part of the year induced by high temperatures and a lack of moisture or lowered environmental humidity.

Extrinsic factor—an influence that affects an organism or community from without.

Fauna—the animals of an area.

Flagging—an effect of wind upon trees when it blows continuously from the same direction for long periods of time; branches, if any, on the windward side of the trunk are stunted, while branches on the lee side are nearly normal in size.

Flora–plants of an area.

Food chain–transfer of food energy from green plants through a series of species.

Food web–complex interlocking patterns of food chains in a biotic community.

Grassland–a treeless, grassy plains area in central parts of temperate North America, Eurasia and Australia.

Habitat–physical place where an organism lives.

Herbicide–chemical that is used to kill plants.

Herbivore–animal that feeds upon plants; a first order consumer.

Heterotroph–organism that feeds upon other organisms.

Hibernation–a state of dormancy in certain animals characterized by lowered body temperatures and decreased metabolic rates.

Homeotherm–warm-blooded animal such as a mammal or bird.

Hydrophyte–aquatic plant.

Immigration–permanent movement of an organism into a new area of residence from another area.

Insecticide–chemical used to kill insects.

Intraspecific–having to do with activities among members of a given species.

Interspecific–having to do with activities among members of different species.

Intertidal zone–area of ocean shore between high and low tides.

Intrinsic factor–an influence that affects an organism or community from within itself.

Isotherm–line drawn on a map connecting points of the same temperature.

Law of the Minimum–ecologic principles proposed by Justus Liebig that states that the growth of a plant is dependent upon the amount of raw materials that is presented to it in least quantity.

Law of Tolerance—an extension of Liebig's Law of the Minimum that states that the absence or failure of an organism can be contolled by the qualitative or quantitative deficiency or excess of any one of several factors that may approach the limits of the requirements for that factor.

Leaching—dissolving out of soluble materials such as salts and minerals by water percolating through soil, especially in areas of high rainfall.

Lichen—plant form composed of two mutualistic organisms, a fungus and an alga.

Life Zone—large geographic areas characterized by relatively uniform climatic conditions and therefore, roughly the same types of plants and animals.

Limiting factor—that aspect of the environment that is responsible for reduced growth or even complete absence of an organism in a given area.

Louisanian Life Zone—area of high and dry ground forested with pine, oak and palmetto with low swampy ground and cypress; absence of low winter temperatures and relatively high rainfall.

Lower Sonoran Life Zone—area of moderate temperatures, low rainfall, cactus, mesquite and creosote bush.

Lunar rhythm—periodic fluctuation that is synchronized with the position of the moon.

Maritime—pertaining to the ocean.

Mesophyte—plant characteristic of soils with average moisture content; neither desert forms nor aquatic forms.

Migration—seasonal movement of organisms to and from a given area.

Monoculture—farming practice in which large quantities of a single crop are grown; characteristic of the United States.

Morphology—the study of the form and structure of organisms.

Mortality—death rate.

Mutualism—symbiotic relationship between two organisms in which both are benefited by the relationship.

Natality—birth rate.

Neap tide—ocean tide in which the lowest high tide occurs; occurs when the effects of the sun and moon tend to cancel each other at the first and third lunar phases.

Niche—the occupation or function of an organism in a community.

Nitrogen cycle—the passage of nitrogen in its various atomic forms as nitrogen gas, nitrates, nitrites, and organic compounds from organism to organism in an ecosystem.

Nocturnal—having to do with the night; pertains to organisms that are active at night.

Parasite—an organism that feeds on another organism (its host) generally without killing it.

Perennial—plant form that lives through several growing seasons.

Permafrost—perpetually frozen ground; found at relatively shallow depths in the Arctic regions where growing season is short and only the upper surface of the ground becomes thawed.

Pesticide — chemical used to kill an organism, plant or animal, that is regarded as a pest.

Physiographic succession—replacement of one community by another as a result of alteration in the topography of a region, usually by erosion or other natural occurrance.

Photoperiod—length of period of illumination; usually relative to length of period of darkness.

Photosynthesis—chemical process occurrring in green plants that utilizes light energy to manufacture high-energy organic compounds from water and carbon dioxide.

Physiology—study of the functions of living organisms and their parts.

Phytoplankton—aquatic plants that float at various depths that are not capable of maintaining a constant position in the face of a water current.

Pioneer species—first organisms to inhabit a newly exposed area.

Plankton—plants or animals that are not capable of maintaining a constant position in the face of a water current.

Poikilotherm–animals that do not maintain a constant body temperature; cold-blooded animals.

Pollution–the introduction of an undesirable change in some aspect of the environment.

Population–all of the members of a particular species occupying a specific space at a particular time.

Precipitation–falling products of atmospheric condensation; rain, snow, hail, etc.

Predator–an organism that kills and eats another organism.

Prevailing westerlies–winds, poleward from the horse latitudes, that have a diagonal movement toward the poles in a somewhat eastward motion.

Prey–an organism that is killed and eaten by another animal.

Primary succession–the replacement of one community by another in which the process is occurring for the first time in that area.

Producer–an organism that is capable of manufacturing its own food from low-energy materials; usually green plants.

Rain shadow–an area on the lee side of a mountain that receives less moisture than the windward side or even possibly even adjacent areas.

Respiration–the chemical breakdown of high-energy organic compounds, with the aid of oxygen, to produce carbon dioxide and water.

Saprophyte–a plant, such as a fungus or mushroom, that feeds on dead and decaying matter.

Secondary succession–the replacement of one community by another in an area that has undergone succession in the past and is presently repeating the process having been interrupted.

Seral stage–one stage in the succession of an area.

Sere–the complex series of changes in an area, from the initial condition to the climax community.

Southern Coniferous Forest–biome of the southeastern United States characterized by various tall pine trees and live oak.

Species–a group of organisms that are capable of interbreeding in nature and producing fertile offspring.

Spring tide—the highest tides, occuring twice monthly, caused by the combined effects of the moon and the sun at the second and fourth quarter lunar phases.

Stomate—small opening in the surface (usually lower surface) of a leaf that allows for the evaporation of moisture from the leaf and the exchange of gases between the leaf and the atmosphere.

Stratification—appearance of several growth layers in a given community, either vertical or horizontal.

Subclimax—a community that appears in a successional sequence that comes prior to the climax stage.

Succession—the orderly replacement of one community by another over a period of time.

Symbiont—an organism that lives together with another one in some relationship with it.

Symbiosis—living together.

Taiga Biome—large geographic area characterized by spruce, fir, and pine forests; typical of the Canadian forests and mountain regions of the United States.

Taxonomy—the systematic classification of organisms.

Topography—the physical appearance of land surfaces.

Transition zone—an area between two biomes or communities, an ecotone.

Transpiration—loss of water from the surface of leaves by evaporation.

Trophic level—feeding level; that level in a food chain in which all organisms of that level obtain their energy in the same number of steps.

Tundra—the treeless plains biome of northern Arctic regions and the higher mountain tops.

Upper Sonoran Life Zone—the short-grass prairies of the grasslands biome.

Virology—the study of viruses.

Water cycle—the passage of water through an ecosystem from lakes, rivers and streams through the atmosphere, back to the earth by precipitation including its temporary side trips through various organisms.

Watershed—that portion of a region that drains into a given lake, stream or river.

Weather—the state of the atmosphere at a particular time.

Xeric—dry, arid.

Xerophyte—plants that grow under relatively arid conditions such as cacti, yucca, mesquite, and creosote bush.

Zoology—the science of animals.

Zooplankton—aquatic animals that float at various depths that are incapable of maintaining a constant position in the face of a water current.

INDEX